TECHNICAL GRAPHICS

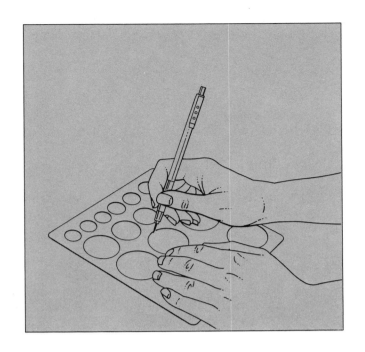

ELECTRONICS WORKTEXT

EDWARD A. MARUGGI

Rochester Institute of Technology

Charles E. Merrill Publishing Company
A Bell & Howell Company
Columbus   Toronto   London   Sydney

To my wife, Carolyn—my strongest advocate in all my endeavors.

Published by
Charles E. Merrill Publishing Company
A Bell & Howell Company
Columbus, Ohio 43216

This book was set in Memphis and Univers.
Cover Designer: Cathy Watterson
Production Editor: Molly Kyle

Cover photo © Photo Researchers, Charles Falco. Motorola MC 68000 16 bit microprocessor. This IC has the equivalent of 70,000 transistors.

International Standard Book Number: 0-675-20311-2
Printed in the United States of America
1 2 3 4 5 6 7 8 9—92 91 90 89 88 87 86

Preface

Technical Graphics: Electronics Worktext presents current methods for producing various technical and electronics drawings required by industry. The electronics drafting fundamentals allow easy transition to a computer-aided drafting (CAD) system. The exercises accommodate students at various levels—high schools, community colleges, technical institutes, vocational programs, industrial training, and continuing and adult education courses. This worktext is relevant to courses in drafting, electronics, electromechanical, electronics, and computer maintenance, and quality control and inspection—in short, any program that requires a basic knowledge of drafting techniques related to electronics.

Each section clearly states learner outcomes and measures performance through practice and review exercises at the end of the section. The many illustrations permit a "self-paced" mode, or the instructor can lecture on individual topics. Tear-out 9" x 12" translucent paper, with format, is provided for drawing work. Advanced drawings can be completed on larger-size vellum.

The worktext reflects the format of a Standard Drafting Practice (SDP) handbook or a Drawing Requirements Manual (DRM), both widely used by industry to standardize drafting procedures. For the most part, the drawing requirements are based on the U.S. government's Department of Defense (DOD), American National Standards Institute (ANSI), and other engineering and professional group practices. Each section is divided into subsections identified by a decimal suffix to the major section number. This format acquaints the student with the style of an SDP handbook or the DRM. The appendixes allow the student to complete the coursework without referring to any additional materials.

I would like to thank the corporations and government agencies that contributed materials to the worktext: Apple Computer, Inc.; Autodesk, Inc.; Beckman Industrial Corp.; CADAM, Inc.; Dale Electronics, Inc.; Dahlgren International; Digital Equipment Company, Inc.; Fairchild Corp.; General Dynamics Corp.; General Instrument Corp.; Harris Corp., R. F. Communications Division; Heath Zenith Computers & Electronics; Hewlett-Packard Corp.; Houston Instruments; International Business Machines Corp.; McGraw-Edison Company, Bussman Division; Motorola, Inc.; Ohmite Manufacturing Company; Olson Manufacturing Company, Inc.; Potter and Brumfield; Reed Devices, Inc.; Signal Transformers Company, Inc.; Sprague Electronics; J. S. Staedtler, Inc.; Summagraphics Corp.; T & W Systems, Inc.; Tektronix, Inc.; Unitrode Corp.; and Wabash Relay & Electronics Company. In addition, I would like to acknowledge the technical assistance and critique of Eder Benati, Professor, NTID @ RIT, Rochester, New York; Anthony Scalise, Harris Corporation, Rochester, New York; Ronald Till, Professor, NTID @ RIT, Rochester, New York; and Albert Luiz, Dahlgren International, Dallas, Texas, who also provided technical illustration assistance.

I also appreciate the suggestions of the technical reviewers provided by the Charles E. Merrill Publishing Company: Ames L. Stewart, of Central Missouri State University; Tim Schmitt, of El Camino College, California; and Earl J. Scribner, of City College of San Francisco.

iii

MERRILL'S INTERNATIONAL SERIES IN ELECTRICAL AND ELECTRONICS TECHNOLOGY

BATESON	Introduction to Control System Technology, Second Edition, 8255–2
BEACH/JUSTICE	DC/AC Circuit Essentials, 20193–4
BERLIN	Experiments in Electronic Devices, 20234–5 The Illustrated Electronics Dictionary, 20451–8
BOGART	Electronic Devices and Circuits, 20317–1
BOGART/BROWN	Experiments for Electronics Devices and Circuits, 20488–7
BOYLESTAD	Introductory Circuit Analysis, Fourth Edition, 9938–2 Student Guide to Accompany Introductory Circuit Analysis, Fourth Edition, 9856–4
BOYLESTAD/KOUSOUROU	Experiments in Circuit Analysis, Fourth Edition 9858–0
BREY	Microprocessor/Hardware Interfacing and Applications, 20158–6
BUCHLA	Digital Experiments: Emphasizing Systems and Design, 20562–X
COX	Digital Experiments: Emphasizing Troubleshooting, 20518–2
FLOYD	Digital Fundamentals, Third Edition, 20517–4 Electronic Devices, 20157–8 Essentials of Electronic Devices, 20062–8 Principles of Electric Circuits, Second Edition, 20402–X Electric Circuits, Electron Flow Version, 20037–7
GAONKAR	Microprocessor Architecture, Programming and Applications with the 8085/8080A, 20159–4
LAMIT/LLOYD	Drafting for Electronics, 20200–0
LAMIT/WAHLER/HIGGINS	Workbook in Drafting for Electronics, 20417–8
MARUGGI	Technical Graphics: Electronics Worktext, 20311–2
NASHELSKY/BOYLESTAD	BASIC Applied to Circuit Analysis, 20161–6
ROSENBLATT/FRIEDMAN	Direct and Alternating Current Machinery, Second Edition, 20160–8
SCHWARTZ	Survey of Electronics, Third Edition, 20162–4
SEIDMAN/WAINTRAUB	Electronics: Devices, Discrete and Integrated Circuits, 8494–6
STANLEY, B. H.	Experiments in Electric Circuits, Second Edition, 20403–8
STANLEY, W. D.	Operational Amplifiers with Linear Integrated Circuits, 20090–3
TOCCI	Fundamentals of Electronics Devices, Third Edition, 9887–4 Electronic Devices, Third Edition, Conventional Flow Version, 20063–6 Fundamentals of Pulse and Digital Circuits, Third Edition, 20033–4 Introduction to Electric Circuit Analysis, Second Edition, 20002–4
WARD	Applied Digital Electronics, 9925–0
YOUNG	Electronic Communication Techniques, 20202–7

MERRILL TITLES IN DRAFTING, MECHANICAL, CIVIL AND GENERAL TECHNOLOGY

BATESON	Introduction to Control System Technology, Second Edition, 8255–2
KEYSER	Materials Science in Engineering, Fourth Edition, 20401–1
LAMIT/LLOYD	Drafting for Electronics, 20200–0
LAMIT/WAHLER/HIGGINS	Workbook in Drafting for Electronics, 20417–8
MARUGGI	Technical Graphics: Electronics Worktext, 20311–2
MOTT	Applied Fluid Mechanics, Second Edition, 8305–2 Machine Elements in Mechanical Design, 20326–0
ROLLE	Introduction to Thermodynamics, Second Edition, 8268–4
WEISMAN	Basic Technical Writing, Fifth Edition, 20288–4

Contents

Section 1
TECHNICAL GRAPHICS PRACTICES 1
1.1   Introduction 2
1.2   Equipment 2
1.3   Tools 4
1.4   Drawing Materials 5
1.5   Lettering and Drafting Techniques 6
1.6   The Freehand Sketch 10
1.7   The Projection Drawing 12
1.8   Dimensioning 14
1.9   Summary 19
1.10 Review Exercises 21
Practice Exercises 171

Section 2
THE BLOCK DIAGRAM 23
2.1   Purpose and Function 24
2.2   Graphic Symbols 24
2.3   Information Flow 24
2.4   Line Convention 24
2.5   Size and Shape of Blocks 25
2.6   Lettering 25
2.7   Method for Drawing the Block Diagram 26
2.8   Summary 26
2.9   Review Exercises 29
Practice Exercises 185

Section 3
THE CONTROL DRAWING 31
3.1   Purpose and Function 32
3.2   Types of Control Drawings 32
3.3   Drawing Preparation 36
3.4   Summary 36
3.5   Review Exercises 41
Practice Exercises 197

Section 4
THE LOGIC DIAGRAM 43
4.1   Purpose and Function 44
4.2   Types of Logic Diagrams 44
4.3   Logic States 44
4.4   Symbol Presentation Techniques 44
4.5   Tagging Lines 46
4.6   Function Identification Letter Combinations 46
4.7   Signal Flow 47
4.8   Method for Drawing the Logic Diagram 47
4.9   Summary 48
4.10 Review Exercises 49
Practice Exercises 207

Section 5
THE SCHEMATIC DIAGRAM 51
5.1   Purpose and Function 52
5.2   Graphic Symbols 52
5.3   Conductor Paths 52
5.4   Reference Designations 54
5.5   Component Values 54
5.6   Method for Drawing the Schematic Diagram 55

5.7   Summary 55
5.8   Review Exercises 57
Practice Exercises 217

Section 6
THE PRINTED CIRCUIT BOARD 61
6.1   Purpose and Function 62
6.2   The Circuit Board Layout 62
6.3   Circuit Board Layout Review 63
6.4   The Artwork 65
6.5   Artwork Drawing Review 66
6.6   The Board Detail Drawing 67
6.7   The Marking Drawing 67
6.8   The Assembly Drawing 67
6.9   Summary 67
6.10 Review Exercises 73
Practice Exercises 233

Section 7
THE INTERCONNECTION DIAGRAM 75
7.1   Purpose and Function 76
7.2   Types of Interconnection Diagrams 76
7.3   Layout of the Diagram 77
7.4   Method for Drawing the Interconnection Diagram 77
7.5   Summary 78
7.6   Review Exercises 79
Practice Exercises 257

Section 8
THE CONNECTION DIAGRAM 81
8.1   Introduction 82
8.2   The Wiring Diagram 82
8.3   The Cable Assembly Diagram 85
8.4   The Wiring Harness Diagram 87
8.5   Summary 89
8.6   Review Exercises 91
Practice Exercises 265

Section 9
ELECTROMECHANICAL PACKAGING 93
9.1   Introduction 94
9.2   Designer/Drafter Responsibilities 94
9.3   Types of Equipment Enclosures 94
9.4   Chassis Layout 96
9.5   Drafting Practices 96
9.6   Fastening Methods 97
9.7   Method for Developing an Electromechanical Package 100
9.8   Summary 100
9.9   Review Exercises 101
Practice Exercises 279

Section 10
INTRODUCTION TO COMPUTER-AIDED DRAFTING SYSTEMS 103
10.1 Introduction 104
10.2 Basic CAD Concept 104

10.3 System Components **105**
10.4 Minicomputer Based Systems **105**
10.5 HP-EGS **109**
10.6 Microcomputer Based Systems **112**
10.7 Summary **118**
10.8 Basic CAD Terminology **118**
10.9 Review Exercises **121**
Practice Exercises **299**

Appendix A
COMMONLY USED GRAPHIC SYMBOLS FOR
ELECTRICAL AND ELECTRONIC DIAGRAMS 123

Appendix B
WORD ABBREVIATIONS ON DRAWINGS 139

Appendix C
CLASS DESIGNATION LETTERS 145

Appendix D
COLOR CODING AND COLOR
ABBREVIATIONS 146

Appendix E
DECIMAL EQUIVALENTS 147
Wire and Sheet Gage **147**
Inch/Metric **148**

Appendix F
OUTLINE DRAWINGS 149

Appendix G
BENDS 159
Component Leads **159**
Radii in Sheet Metals **160**
90-Degree Developed Length **161**

Appendix H
CHARTS AND CONVERSION TABLES 162
Drill Size Chart **162**
Standard Screw Thread Chart **163**
Trigonometry Chart **165**
Miscellaneous Conversion Tables **168**

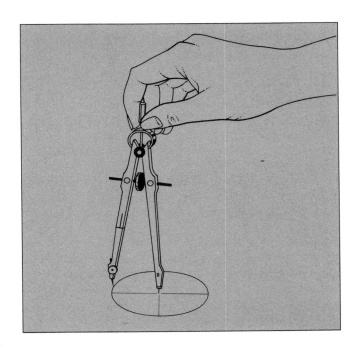

Section 1

TECHNICAL GRAPHICS PRACTICES

LEARNER OUTCOMES

The student will be able to:

- Identify four major pieces of equipment used in a drafting room

- List six different drafting tools and their uses

- Describe two types of drawing materials

- List the various standard sizes of drawing materials

- Identify ten kinds of lines used on electronic and electromechanical drawings

- Identify the most frequently used freehand sketches

- Explain the concept of orthographic projection

- State six fundamental rules for dimensioning

- Demonstrate learning through successful completion of practice and review exercises

1.1 INTRODUCTION

Technical graphics is a method of communication and the language of industry. It is a process of communicating visually and graphically what takes the place of many words. This communication occurs in the engineering department and in the manufacturing facility, between contractor and customer, between buyer and vendor.

Designing a product from initial concept to final assembly requires the energy and talent of various technical and professional personnel. One of the key people in the process is a drafter, who takes a design and, through the skill of technical graphics and knowledge of its practices and procedures, completes the task so that parts, subassemblies, and assemblies can be produced.

The electronics industry requires many types of drawings and diagrams for the several applications and uses of consumer, commercial, and military products that are produced. Some types of drawings are required more than others depending upon the nature of the industry's product or product line. This worktext identifies the most important and most frequently used electronic diagrams and drawings required by goods-producing industries. They include:

- The block diagram
- The component drawing
- The logic diagram
- The schematic diagram
- The printed circuit board and its associated drawings
- The interconnection diagram
- The connection diagram
- Electromechanical packaging
- Introduction to computer-aided drafting (CAD) systems

The practices in this worktext are those that have become standardized over the years and are universally accepted by industrial contractors of military products as well as by large and medium-sized companies producing consumer and commercial products.

This section provides experiences in basic technical graphics as a foundation, so that all subsequent sections in this worktext can be completed successfully. These practices are required for students who are pursuing a program in an electronics or a related field. For greater knowledge of mechanical or industrial drafting, please refer to one of several drafting textbooks available in libraries and bookstores.

In this worktext, we will sometimes use the terms *part, subassembly, assembly, unit,* and *system* as divisions of equipment. For the sake of clarity, we will define them here:

- **Part** A part is one piece, or two pieces joined together, that is not normally subject to disassembly without destroying its designed use. (Examples: composition resistor, gear, cathode ray tube.)
- **Subassembly** A subassembly consists of two or more parts that form a portion of an assembly or a unit replaceable as a whole, but having parts that are individually replaceable. (Examples: telephone dial, gun mount stand, window sash.)
- **Assembly** An assembly consists of a number of parts or subassemblies or any combination thereof, joined together to perform a specific function. (Examples: power supply, pre-amplifier.)
- **Unit** A unit is a major portion of a set or system. It consists of a collection of parts, subassemblies, and assemblies packaged together as a physical entity. (Examples: a rack of electronic equipment, computer console.)
- **System** A system is a combination of units, assemblies, and parts, necessary for performing an operational function or functions. (Examples: a complete test station, telephone carrier system, or facsimile transmission system.)

Equipment division definitions are illustrated in Figure 1.1.

1.2 EQUIPMENT

Technical drawing can be performed with a minimal investment in equipment and basic tools, or one can spend large sums of money to outfit the drafting room. Modern drafting rooms are equipped with large, full-size drafting stations that include drafting boards and drafting machines designed to move in any direction

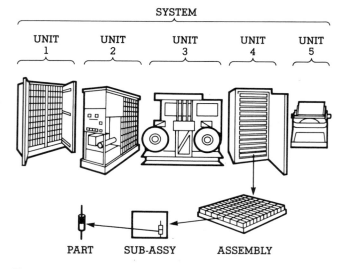

Figure 1.1
Illustration of equipment divisions.

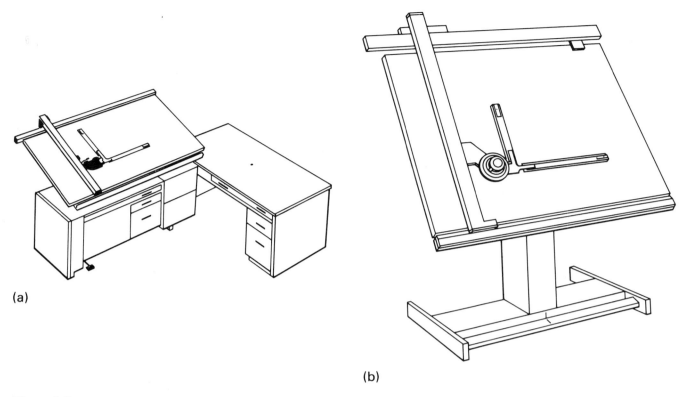

Figure 1.2
(a) Modern drafting station; (b) Drafting machine on a drafting table.

quickly and accurately to accommodate the technical needs of the drafter. In addition, drafting chairs are mobile and adjustable to discrete heights. The drafting station usually includes a reference table located behind or to the side of the drafter. This table is used for reference material the drafter needs for his work and may include drawings, layouts, drafting manuals, reference books, and catalogs.

Drafting machines are normally mounted on the top edge of the drafting board and are available to accommodate both left-handed or right-handed drafters. Because of their mobility, these machines allow for rapid and accurate drawing of horizontal, vertical and angled lines with minimum effort by the drafter. Figure

1.2(a) illustrates a typical drafting station in a modern drafting room with (b) a drafting machine mounted to a drafting board.

For technical or financial reasons, the T-square and/or the parallel straightedge is sometimes used instead of the drafting machine to produce technical drawings. Figure 1.3 shows the physical appearance of both the T-square and the parallel straightedge.

The drafting machine, T-square, and parallel straightedge all can be used to produce electronic diagrams and drawings successfully. Because T-squares and straightedges can only be used to draw horizontal lines, however, triangles must be used for drawing vertical or angled lines.

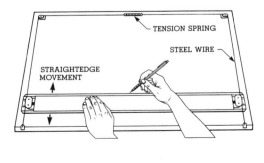

(a)

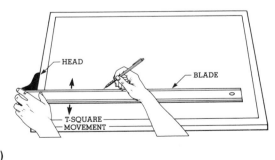

(b)

Figure 1.3
(a) Parallel straightedge; (b) T-square.

1.3 TOOLS

When producing electronic-type technical diagrams or drawings, several special tools of the trade are required.

1.3.1 TRIANGLES

The drafter normally uses three types of triangles: the 45-degree, the 30/60-degree, and the adjustable triangle, shown in Figure 1.4. Used individually, the 45-degree and 30/60-degree triangles only allow one to draw lines at 30 degrees, 45 degrees, 60 degrees, and 90 degrees from the horizontal. When used together, however, they can produce additional angles of 15 degrees and 75 degrees, as shown in Figure 1.5.

Figure 1.6(a) shows how to draw horizontal lines using a parallel straightedge or a T-square, and Figure 1.6(b) shows the proper method of drawing angled lines using a triangle.

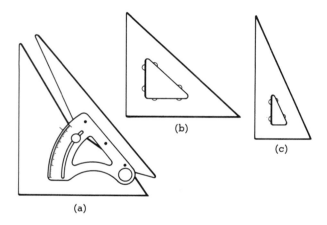

(a)

Figure 1.4
(a) Adjustable triangle; (b) 45-degree triangle; (c) 30/60-degree triangle.

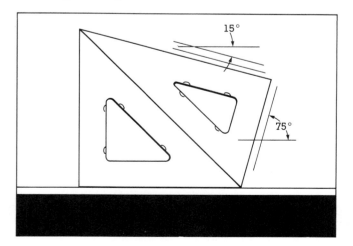

Figure 1.5
Drawing 15-degree and 75-degree angles using two triangles with a straightedge or T-square.

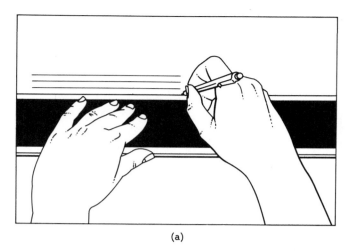

(a)

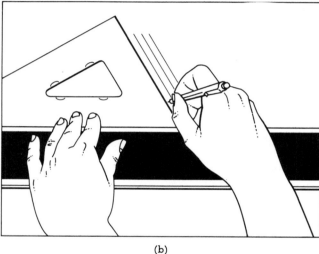

(b)

Figure 1.6
(a) Horizontal lines, using a T-square or a straightedge; (b) Proper method for drawing angled lines.

1.3.2 COMPASSES, TEMPLATES, AND CURVES

Some other tools are required for drawing various special shapes. The compass is used to draw arcs and circles, as demonstrated in Figure 1.7. Templates are widely used by drafting room personnel and include several types. A template is a timesaving device because it allows for the rapid and accurate drawing of shapes that are difficult to produce by other methods or tools. Many types of templates are available specifically for use in producing electronic diagrams and drawings, including one that complies with the American National Standards Institute (ANSI) Y32.2, Graphic Symbols for Electrical and Electronics diagrams. A comprehensive list of electronics symbols are identified in Appendix A, page 123. Sections 4, 5, 6, 8, and 9 have exercises designed for using these templates. Examples of available templates are shown in Figure 1.8.

When using a template, one should be careful to

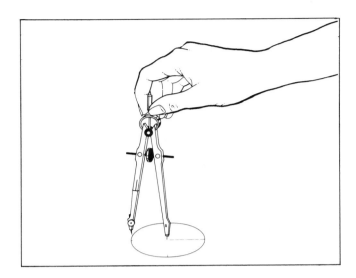

Figure 1.7
How to use a compass.

should represent a smooth curve.

1.4 DRAWING MATERIALS

Materials for producing diagrams and drawings are available in paper, cloth, or a film base. The two types of material most frequently used for drafting purposes include *vellum,* a white, translucent rag content paper, and a film that is durable, dimensionally stable, and can be stored for many years with no degradation of drawing quality. The film type is available under several commercial trademark names. In addition, the advent of copiers that use a toner has meant that 8½″ × 11″, 8½″ × 14″ or larger sketches and drawings can be re-

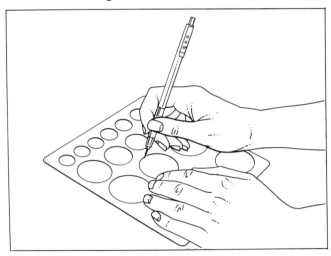

Figure 1.9
How to use a circle template.

lift the template after drawing each character or symbol. Do not slide the template along the drawing surface. When inking, wait for the ink to dry before lifting or moving the template. Figure 1.9 shows the use of templates.

Curves are used to draw irregular shapes. Irregular or noncircular shapes are frequently used in producing development or isometric drawings. The shape is determined by a series of plotted point locations. The curve is then used to connect the plotted points with a line, as shown in Figure 1.10. The completed line

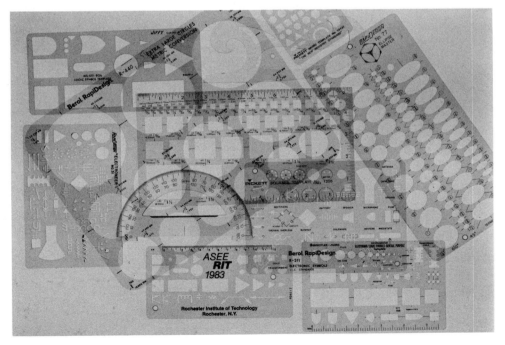

Figure 1.8
Examples of templates.

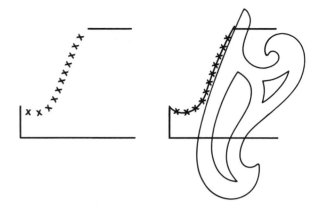

Figure 1.10
Curved shape drawn with a French curve.

produced from plain lined or unlined bond and from reproducible or nonreproducible grid paper. These copiers allow engineering personnel to make available for the drafter's use reductions, enlargements, or copies of parts of diagrams. Drawing materials are available in standard sizes, either plain or with the company's format preprinted on one side. (The practice exercises in the back of this worktext represent a printed format on translucent bond.)

Table 1.1 represents frequently used drawing sizes that have become standardized and are in accordance with MIL-STD-100A, a military drawing standard for those industries under contract to the U.S. Government. Normally, the use of roll-size drawing material is restricted to schematic, logic, connection, and interconnection diagrams that can sometimes be very large because of the complexity of the equipment involved.

1.5 LETTERING AND DRAFTING TECHNIQUES

Two methods of lettering are available to the electronics drafter. One method is called *freehand lettering;* the other is a mechanical one accomplished with a lettering guide. All lettering on a diagram or drawing must be of high quality and legible regardless of the final reduction size requirement of the drawing. Some people have a natural talent for lettering, while others need to work hard to become proficient. Practice is the key.

1.5.1 STYLES OF LETTERING

Of the several styles of lettering that have been developed over the years, the predominant one for electronic drawings is the single-stroke, uppercase, commercial Gothic style. *Single stroke* means that the required thickness or weight of each letter is formed using one stroke; *uppercase* indicates that all letters are capitalized and *gothic style* is one in which all strokes of each letter are even. Either vertical or inclined letters are acceptable in industry, but most large companies prefer vertical characters. Only one type of lettering, either vertical or inclined, should appear on a drawing. Methods for producing both vertical and inclined letters and numerals are illustrated in Figure 1.11.

Notes, either general or localized, are placed on a diagram or drawing to align parallel to the bottom of the drawing. Fractions and mixed numbers are drawn with a horizontal bar, as shown in Figure 1.12.

1.5.2 SPACING AND SIZES

Photographically reducing original drawings to a smaller size or reproducing them on microfilm limits the minimum height of characters and line spacing in terms of maintaining legibility. The success of either reproduction process depends on clear, legible, well-formed letters and numerals. Care should be taken to produce open letters and numerals with sufficient space between them to assure legibility after reduction. Spacing between words should not be less than the height of one letter and spacing between lines should not be less than the height of letters. In all the exercises in this worktext, it is recommended that minimum letter and numeral heights be in accordance with Table 1.2.

Table 1.1
Frequently used drawing sizes.

SIZE (IN INCHES)		LETTER SIZE
WIDTH	LENGTH	
11	x 8½	A
11	x 17	B
17	x 22	C
22	x 34	D
34	x 44	E
28	x 40	F
28	x 144 max	H (ROLL)
34	x 144 max	J (ROLL)

Table 1.2
Height of lettering.

LOCATION OF LETTERING	NUMERAL AND LETTER HEIGHT IN INCHES
DRAWING TITLE	.18 HIGH
DRAFTER'S NAME, ALL LETTERING ON FACE OF DRAWINGS, DIMENSIONS, NOTES, TOLERANCES, AND HOLE CHARTS	.12 HIGH

VERTICAL INCLINED

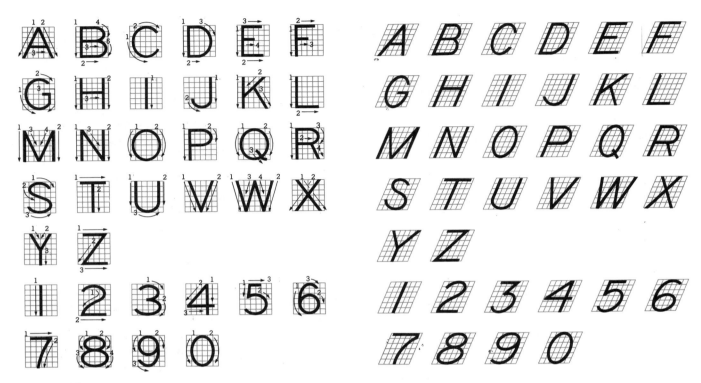

Figure 1.11
Method for producing gothic style lettering.

FRACTIONS			MIXED NUMBERS	
$\dfrac{3}{4}$	$\dfrac{1}{2}$	$\dfrac{5}{8}$	$1\dfrac{3}{16}$	$4\dfrac{5}{32}$

Figure 1.12
Acceptable form for fractions and mixed numbers.

Figure 1.13
Ames Lettering Guide.

1.5.3 LETTERING AIDS

Lettering aids help assure that lettering on a drawing is of good quality by providing uniformity to height and width of characters. When using freehand lettering, it is necessary to begin with horizontal, vertical, and/or inclined guide lines. An example of a commercially available template for drawing guidelines is the Ames Lettering Guide shown in Figure 1.13. This lettering aid

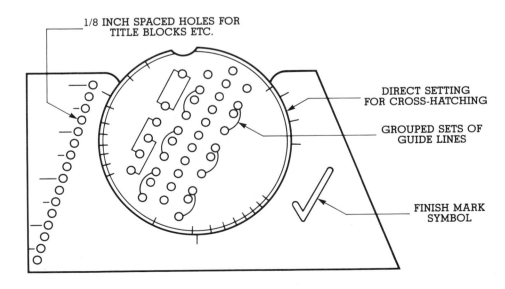

provides guidelines as illustrated in Figure 1.14. Figure 1.15 shows how to use the guide in conjunction with a T-square, drafting machine, or straightedge. Guidelines should be light enough that they will not reproduce when the drawing is copied.

The Ames lettering guide is used primarily to provide spacing or lines for characters. The lettering templates in Figure 1.16 are used to actually draw letters and numerals. These templates are made of various thicknesses of plastic and are available in several character sizes and styles for either vertical or angled lettering.

Figure 1.15
Use of an Ames lettering guide.

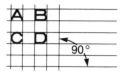 

Figure 1.14
Vertical, angled, and horizontal guidelines.

Figure 1.16
Lettering templates.

1.5.4 LINE CONVENTION

Several different kinds and widths of lines are used for electronics diagrams and drawings. The four major line widths are thin, medium, thick, and extra-thick. All lines should be opaque and of uniform width throughout their length. Line characteristics are illustrated in Figure 1.17 and are described as follows:

■ **Center line** A thin line made up of long and short dashes, alternately spaced and consistent in length, beginning and ending with a long dash. Center lines cross each other without voids. Very short center lines may be unbroken if there is no possibility of confusion with other lines. (*Use:* To indicate the axis of a part or feature, paths of motion, and to indicate the theoretical line about which a part or feature is symmetrical.)

■ **Dimension Line** A thin line the same width as center line, terminating with arrowheads at each end and unbroken except where the dimension is placed. (*Use:* To indicate the extent and direction of dimensions.)

■ **Leader line** A thin line the same width as a center line, terminating in an arrowhead or dot. Arrowheads should always terminate at a line. Generally, a leader line should be an oblique straight line except for a short horizontal portion extending to mid-height, preferably to the first or last letter or digit of the note. (*Use:* To indicate a part or portion of a drawing in which a number, note, or reference applies.)

■ **Break line** A thin line and freehand zigzags for long breaks; thick freehand line for short breaks. (*Use:* To show an area or a portion of a part that has been removed to show hidden detail, to limit a partial section or view, and to eliminate repeated detail.)

■ **Extension line** A thin continuous line that does not touch the outline. (*Use:* To extend points or planes to indicate dimensional limits.)

■ **Phantom line** A medium line consisting of one long and two short dashes, alternately and evenly spaced, with a long dash at each end. (*Use:* To show alternate positions of parts, relative position of adjacent parts shown for reference, and to eliminate repeated detail.)

■ **Section line** A thin continuous line usually drawn at a 45-degree angle in adjacent parts. (*Use:* To indicate surfaces exposed by a section cut.)

■ **Hidden line** A medium-weight line of short dashes, closely and evenly spaced. It is less prominent than an outline or visible lines. Hidden lines begin and end with a dash in contact with the line from which they start and end. (*Use:* To show hidden features of a part.)

■ **Outline or visible line** A thick line, usually the most prominent one on a drawing. (*Use:* For all lines on the drawing representing visible lines of an object.)

■ **Datum line** A medium line consisting of one long dash and two short dashes evenly spaced. (*Use:* To indicate the position of a datum plane.)

■ **Cutting plane** and **Viewing plane lines** Thick lines slightly heavier than outline or visible lines. Arrowheads, drawn at 90 degrees to the cutting or viewing plane line, indicate the viewing directions. (*Use:* Cutting plane lines indicate a plane or planes in which a section is taken; viewing plane lines indicate the plane or planes from which a surface or surfaces are viewed.)

Some examples of line conventions appear in Figure 1.18.

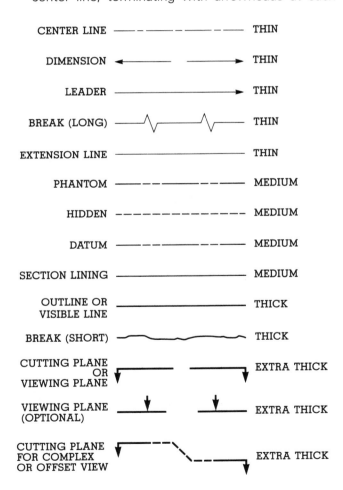

Figure 1.17
Line characteristics.

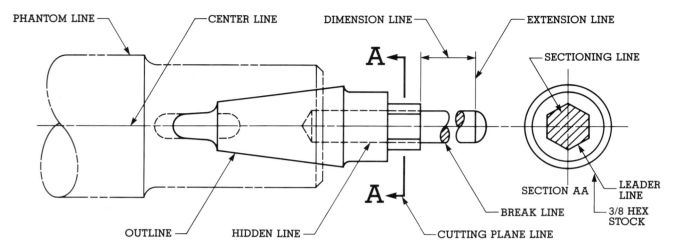

Figure 1.18
Examples of line conventions.

1.5.5 LEADS AND HOLDERS

Leads and mechanical lead holders have, for the most part, replaced the pencil for drafting purposes. Several different grades of lead are available ranging from very soft to very hard. Table 1.3 identifies the most popular leads and their uses. Leads are graded so that the higher the "H" number, the harder the lead. Table 1.3 is only a guide; drafters have personal preferences as to which leads produce the best quality.

Leads cannot be used without the mechanical lead holder. Lead holders or mechanical pencils, as they are sometimes called, are available in many types and vary according to lead diameter and type of lead. Figure 1.19 shows two of the most common holders.

1.6 THE FREEHAND SKETCH

Knowing how to interpret and draw the freehand sketch is an important skill for electronics drafters. The freehand sketch is often the primary, and sometimes the only, communication message between the engineering project leader and the drafter. Also, it may well be the first drawing the drafter will be required to produce on a project.

The drafter is not the only person who produces a freehand sketch. Sketches are drawn by engineers who design and develop circuitry, schematics, and logic diagrams. Often model shop personnel, instrument makers, and engineering model builders draw sketches as part of their work. Production engineers use sketches to show initial thoughts on a design. In short, anyone involved with the engineering and production of goods and services whose communication needs cannot be met by the written or spoken word use and benefit from the freehand sketch.

Many preliminary concepts and ideas take the form of a sketch to convey data that may later become a finished diagram or drawing. Technical freehand sketches, although infrequently drawn to scale, are usually drawn to correct proportions. The drafter should attempt to produce quality sketches because at this point changes can be made quickly. Planning for the final drawing includes determining placement of

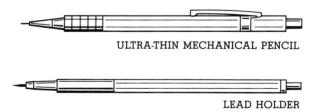

ULTRA-THIN MECHANICAL PENCIL

LEAD HOLDER

Figure 1.19
Lead holders.

Table 1.3
Leads and their uses.

LEAD HARDNESS	USAGE
6H 5H 4H 3H 2H H	GUIDE LINES, CONSTRUCTION LINES CENTER LINES MEDIUM WEIGHT USES AND FOR GENERAL DRAWING THICK LINES, LETTERING SKETCHING

views, dimensions, location of symbols, positioning of notes, hole-size charts, and identifying the size of drawing material.

1.6.1 TYPES OF SKETCHES

Several types of sketches are used in designing, developing, and drafting electronic and electromechanical equipment. The most frequent freehand technical sketches include the multiview drawing and the isometric drawing. An example of a simple multiview sketch is illustrated in Figure 1.20. The same part that is shown in Figure 1.20 is drawn as an isometric sketch in Figure 1.21. Note that isometric lines are sketched parallel to a 30-degree isometric axis. Vertical and horizontal measurements are scaled directly from the orthographic drawing and transferred to the corresponding isometric lines.

The only materials required for sketching include a soft grade of lead, a mechanical lead holder or pencil, an eraser, and paper. The most suitable paper is either ruled with a reproducible or nonreproducible grid of 4, 8, 10, or 12 squares to the inch, or a plain translucent paper with no grid used in conjunction with a grid master underneath to provide grid lines.

1.6.2 SUGGESTIONS FOR PRODUCING A FREEHAND SKETCH

Drafting styles and the quality of drawing vary from person to person, depending mostly upon experience.

These suggestions will help you produce quality technical freehand sketches.

1. Select drawing materials based on the size, complexity, and reproducibility requirements of the sketch.
2. Determine the type of sketch required. For example, will it be multiview, isometric, an elementary electrical sketch, a logic or connection diagram? Will it require dimensions and notes?
3. Do not attach the drawing paper to the drawing surface so you will have the option of placing the paper at any angle that is comfortable. Lines can be drawn smoothly, accurately, and straight if as much of the arm as possible is supported by the drafting board surface.
4. For most people, sketching horizontal lines is easier than drawing vertical or angled ones. Therefore, try to make certain that most, if not all, of the lines are drawn horizontally, as illustrated in Figure 1.22. You may have to rotate the drawing several times during the production of the sketch.
5. Although sketching circles or arcs can be particularly difficult, the method described in Figure 1.23 will help. The method calls for drawing center lines first, then determining the circle size in your mind and translating it into equally spaced radial points or lines (at least four or as many as twelve; more radial lines will assure a more perfect, smoother circle). Finally, convert the radial points with an even constant pressure of the lead on the paper. The same procedure may be used for sketching arcs.

TOP VIEW

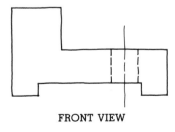
FRONT VIEW

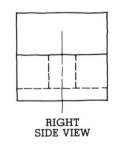
RIGHT SIDE VIEW

Figure 1.20
Multiview freehand sketch.

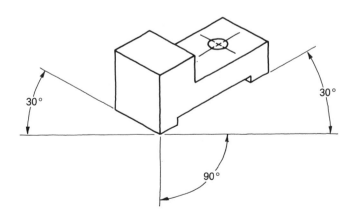

Figure 1.21
Isometric freehand sketch.

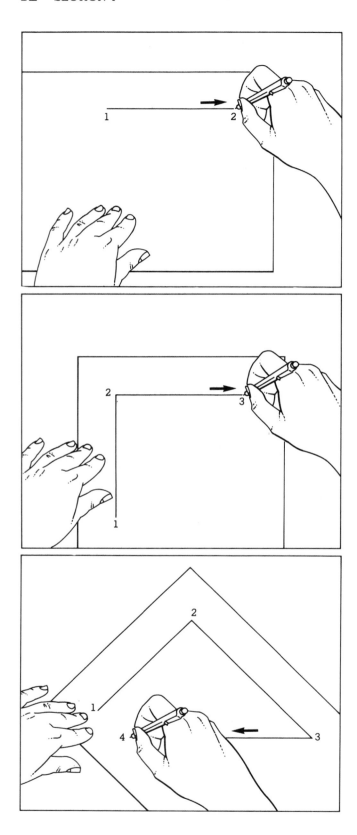

Figure 1.22
Sketching lines.

1.7 THE PROJECTION DRAWING

The three major drawing methods for preparing electronic and electromechanical graphics are the orthographic projection, the isometric projection, and the perspective projection. The orthographic projection is the preferred and standard type of drawing.

1.7.1 ORTHOGRAPHIC PROJECTION

To understand orthographic projection, visualize the transparent box illustrated in Figure 1.24, where the object to be drawn is placed inside the box so that its axes are parallel to the axes of the box. The projection on the sides of the box are the views you see by looking straight at the object through each side in turn. If each view is drawn as seen on the side of the box and the box is unfolded and laid flat (as shown in Figure 1.25), the result is a six-view orthographic projection drawing. Any of the six views could be regarded as the principal view, and the box could be folded out from one view as easily as another. When the box is unfolded from the front view of the object as indicated in Figure 1.25(b), it is called *third-angle orthographic projection.* When trying to determine how many views to draw, remember to draw only those views that most clearly convey the size and contour of the object with the least number of hidden lines.

1.7.2 ONE-VIEW DRAWINGS

One-view drawings are acceptable for items such as cylinders, spheres, or square parts if the necessary dimensions are properly indicated. Thin objects of uni-

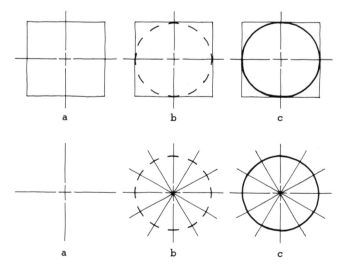

Figure 1.23
Sketching circles.

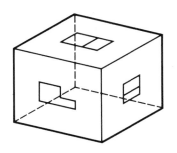

Figure 1.24
Transparent box.

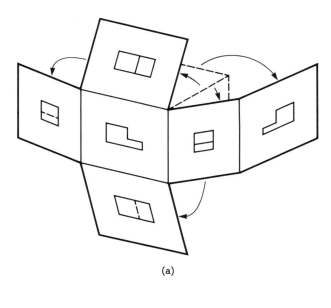

(a)

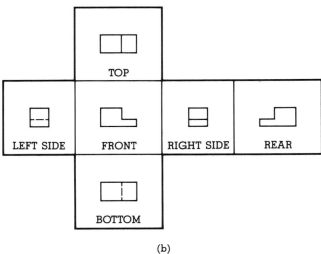

(b)

Figure 1.25
(a) Unfolding box; (b) Six-view orthographic projection.

form thickness, such as shims, gaskets, and plates, may also be shown by a single view as long as a note on the drawing indicates the material thickness. (See Figure 1.26 for examples of one-view drawings.)

1.7.3. TWO-VIEW DRAWINGS

Two-view drawings may be arranged as any two adjacent views (views next to one another), as shown in Figure 1.25(b). Two examples of two-view drawings are illustrated in Figure 1.27.

1.7.4 THREE-VIEW DRAWINGS

Three-view drawings may be arranged as any three adjacent views in the relation shown in Figure 1.25(b).

The example in Figure 1.28(a) shows the front, top, and the right side view of a mounting bracket; Figure 1.28(b) illustrates the front, left, and right side views. To emphasize a point made earlier, you need to draw only those views that clearly outline the shape and size of the object.

1.7.5 AUXILIARY VIEWS

Sometimes objects that have inclined surfaces or faces, or other features that are not parallel to any of

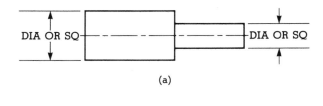

(a)

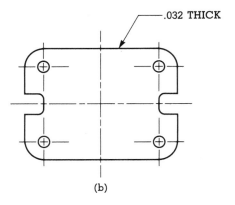

(b)

Figure 1.26
One-view drawings.

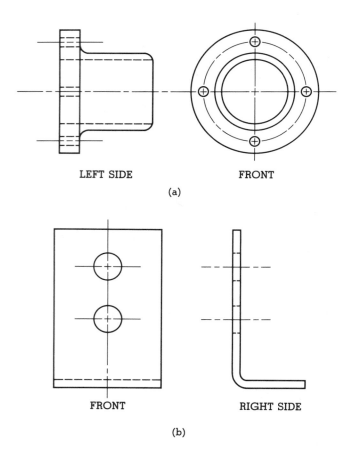

LEFT SIDE   FRONT

(a)

FRONT   RIGHT SIDE

(b)

Figure 1.27
Two-view drawings.

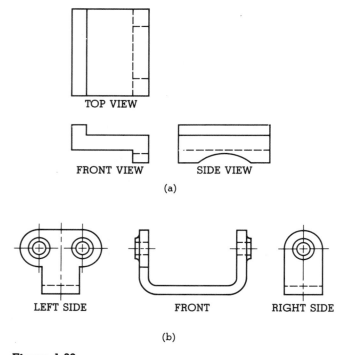

TOP VIEW

FRONT VIEW   SIDE VIEW

(a)

LEFT SIDE   FRONT   RIGHT SIDE

(b)

Figure 1.28
Three-view drawings.

the three principal planes of projection, may require auxiliary views to show their true shape. Auxiliary views are developed as though the auxiliary plane were hinged to the object to which it is perpendicular and then revolved into the plane of the paper, as indicated in Figure 1.29. There are several kinds of auxiliary views, including the partial auxiliary view and the left and right side auxiliary views, as portrayed in Figure 1.30.

1.8 DIMENSIONING

Few, if any, dimensions are required for most electronic diagrams. In a mechanical or electromechanical drawing, however, dimensions must be shown if the item is to be produced. A dimension is a numerical value expressed by appropriate units of measurement and indicated on the drawing by lines, symbols, and notes to define the object's geometrical characteristics. The preferred method of dimensioning is that shown in American National Standards Institute (ANSI) standard ANSI Y14.5 (1982), "Dimensioning and Tolerancing for Engineering Drawings," illustrated in Figure 1.31. All dimensions are read in alignment with the bottom and left edge of the drawing and referred to as the unidirectional method or datum-line method of dimensioning.

The dimensioning practices outlined in this section are basic and do not reflect the various symbols used in the American National Standards Institute (ANSI) 1982 specification.

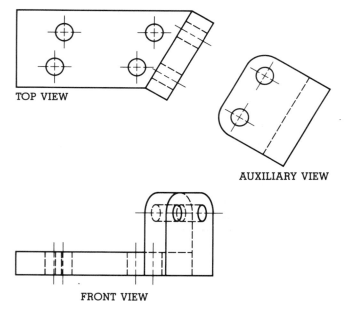

TOP VIEW

AUXILIARY VIEW

FRONT VIEW

Figure 1.29
Auxiliary view.

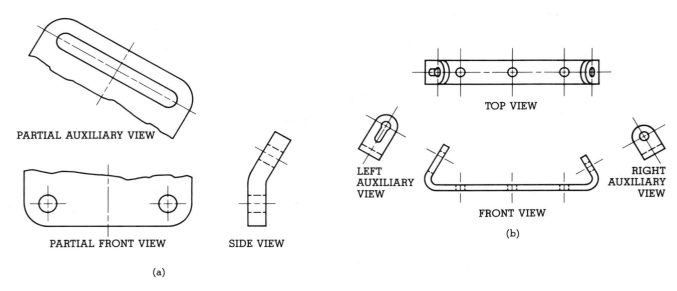

Figure 1.30
Partial, left, and right side auxiliary views.

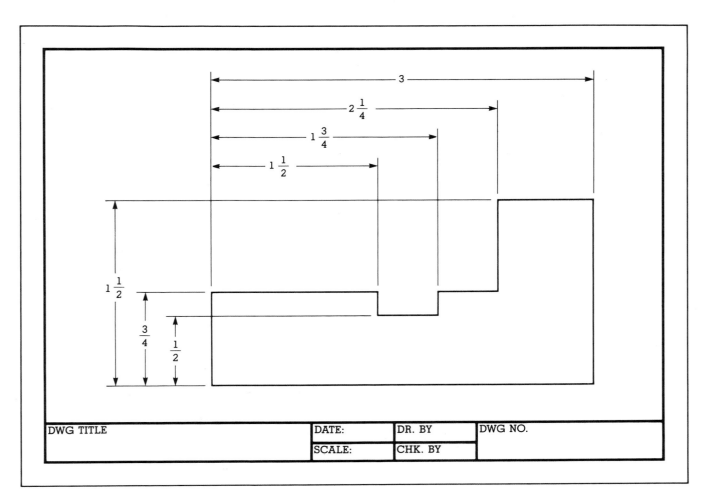

Figure 1.31
Unidirectional or datum-line method of dimensioning.

1.8.1 FUNDAMENTAL RULES

For dimensions to define the geometrical characteristics clearly and concisely, you must follow a few basic rules to produce quality graphics.

- Show enough dimensions so that the intended sizes and shapes can be determined without calculating or assuming distances.
- State each dimension clearly, so it can be interpreted in only one way.
- Show the dimensions between points, lines, or surfaces that have a necessary and specific relation to each other or that control the location of other components or mating parts.
- Dimension, extension, and leader lines shall not cross each other unless absolutely necessary.
- Select and arrange dimensions to avoid the accumulation of tolerances.

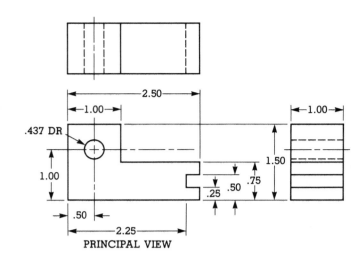

Figure 1.32
Three-view drawing.

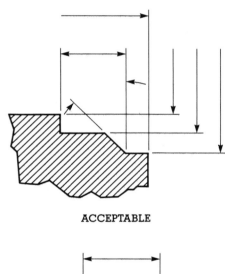

ACCEPTABLE

UNACCEPTABLE

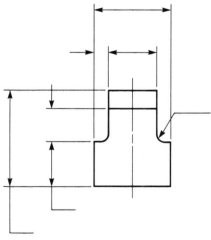

ACCEPTABLE

UNACCEPTABLE

Figure 1.33
Acceptable and unacceptable

- Center lines, object lines, or extension lines should not be used as dimension lines.
- Show each dimension only once.
- Whenever possible, dimension each feature in the view where it appears. Dimensioning to hidden lines should be avoided.
- Unless clarity is improved, dimensions are shown outside the outline of the part.

1.8.2 APPLICATION OF DIMENSIONS

To apply dimensions, one must first determine which view is the principal view of the object. The principal view is the one that most completely shows the object's characteristic shape. Show as many dimensions as practical in the principal view, but avoid overcrowding. All dimensions for surfaces that show in this profile should be given in this view. Figure 1.32 illustrates a three-view drawing with the principal view showing the most dimensions.

1.8.2.1 Crossing Lines

Avoid crossing lines on a drawing wherever possible. To avoid crossing, the dimension for the shortest length of the object should normally be placed nearest the outline of the object and adjacent parallel dimension lines should be added in order of their size, with the longest dimension line the outermost one.

Dimension lines should not be broken when they cross extension lines or leader lines. If extension lines cross dimension lines close to arrowheads, however, a break in the extension line is recommended. Figure 1.33 shows acceptable and unacceptable methods for crossing lines on a drawing.

1.8.2.2 Grouping Dimensions

In grouping dimensions, clarity is improved by placing dimension lines, extension lines, and numerals in line where space permits, as indicated in Figure 1.34.

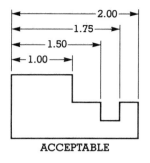
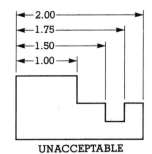

ACCEPTABLE   UNACCEPTABLE

Figure 1.35
Staggering dimensions.

1.8.2.3 Staggering Dimensions

When space becomes restricted on a drawing, it is better to stagger columns of dimensions to eliminate the possible interference of numerals. (See Figure 1.35.)

1.8.2.4 Placement of Dimensions

Wherever practical, dimensions should be placed outside the view being dimensioned. In general, the overall dimensions of a part are placed above and on the right of the principal view. Overall dimensions of surfaces that show in profile in two views are placed between these views, as in Figure 1.36. Dimensions are placed between extension lines wherever possible, as illustrated in Figure 1.37.

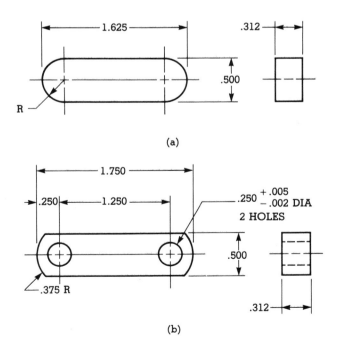

Figure 1.36
Placement of dimensions.

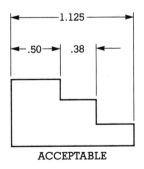
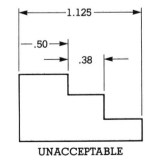

ACCEPTABLE   UNACCEPTABLE

Figure 1.34
Grouping dimensions.

ACCEPTABLE UNACCEPTABLE

Figure 1.37
Use of extension lines.

1.8.2.5 Dimensions for Holes

Specific dimensioning information pertaining to round holes is illustrated in Figure 1.38, and normally shows the size and the diameter tolerance of the hole. Figure 1.38(a) shows a leader drawn to the outside point of the hole but in line with the centerline of the hole. When a general tolerance note is not shown on a drawing, the hole must be provided a tolerance as given in Figure 1.38(a)(b)(c) and (d). In addition, when hole diameters are indicated by dimension lines as in Figure 1.38(b)(c) and (d), the abbreviation for diameter (DIA) does not follow the hole size. On the drawing, holes are normally located by dimensions to extensions of center lines.

1.8.2.6 Dimensioning Radii

Curved surfaces shown on arcs of circles are dimensioned by drawing a radial dimension through the origin of the radius to the surface in question. The radial dimension line on small radii may be drawn on the side opposite the center instead of through it. The letter *R* follows the dimension of the radius. The preferred placement of the dimension for various sizes and types of radii is shown in Figure 1.39.

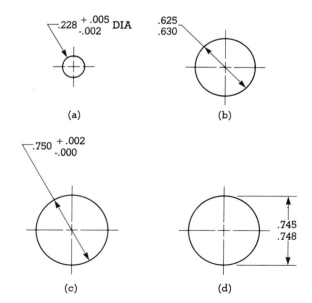

Figure 1.38
Dimensioning of holes.

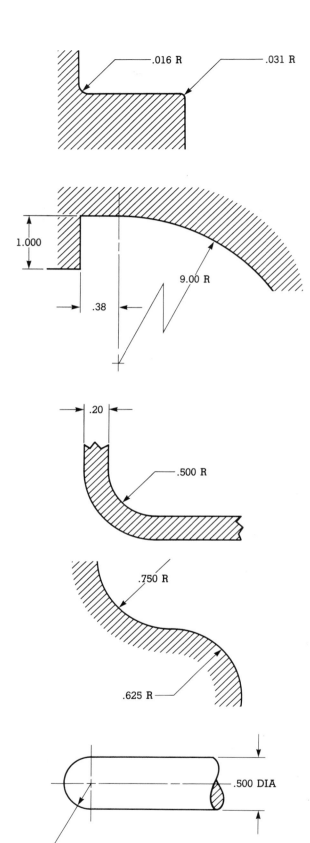

Figure 1.39
Preferred radii dimensioning.

1.9 SUMMARY

Skill in basic technical graphics practices is required for success in the area of electronics drafting. Knowledge in the use of tools and equipment, and proficiency in lettering quality, quality of lines, dimensioning practices, freehand sketching, and projection drawing are all necessary prerequisites to operating a computer-aided drafting (CAD) system. Section 1 has explained several technical graphics practices and identified procedures and rules for producing quality work. Practice exercises for this section are on pages 171 through 183.

1.10   REVIEW EXERCISES

1.   What is technical graphics?

2.   How does the function of a T-square and parallel straightedge differ from that of a drafting machine?

3.   What three types of triangles do drafters use?

4.   Why is the template a valuable tool?

5.   Describe two types of drawing materials used in electronics drafting.

6.   What is the standard drawing paper size for an ''A,'' ''C,'' and ''E'' drawing?

7.   Describe two methods for lettering an electronic drawing.

8.   What is the predominant style of lettering used on electronics drawings?

9.   What are the correct names for the following conventional lines?

   a. ⬅———— ————➡   ————

   b. ——— – ——— – ——— – ———   ————

   c. — — — — — — — — — — —   ————

   d. ▬▬▬▬▬▬▬▬▬   ————

e.

10. Is a 4H lead harder than a 6H? A 2H softer than an H?

11. Give two examples of objects that require a one-view drawing.

12. What are the typical views required for a three-view drawing?

13. Define *dimension*.

14. List six fundamental rules for dimensioning.

15. Which view is the principal view on a drawing?

16. Explain the concept of orthographic projection.

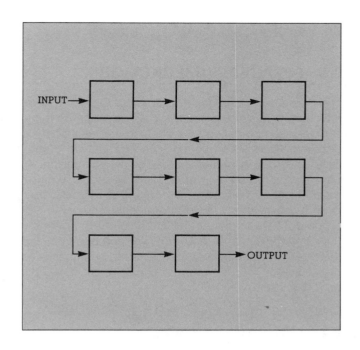

Section 2

THE BLOCK DIAGRAM

LEARNER OUTCOMES

The student will be able to:

- Identify and draw several standard block diagram symbols

- Lay out a simple block diagram showing conventional and auxiliary flow path lines

- Complete a block diagram, given partial information

- Produce a block diagram, given a freehand rough sketch

- Demonstrate learning through successful completion of practice and review exercises

2.1 PURPOSE AND FUNCTION

The purpose of the block diagram is to represent complex units of a system or subsystem of equipment in a simple, clearly outlined sequence so that relationships between various components can be viewed graphically. This representation is used as an aid in planning, designing, drafting, and fabricating electronic, electromechanical, or mechanical equipment. Functionally, each block of the diagram identifies a component or group of components, their location in the system, the flow of information to and from the components, and the title that identifies the components. Additional data often include inputs, outputs, control point locations, and data paths.

2.2 GRAPHIC SYMBOLS

A rectangle or a square is the basic symbol for identifying a functional unit on a block diagram. Sometimes triangles and circles are also used. The functional unit is represented by other shapes as well, including electronic symbols, as on the schematic diagram. Common electronic symbols on a block diagram include those for the capacitor, resistor, speaker, antenna, switch, battery, connector, and meter. In addition, inputs and outputs are often shown as electronic symbols, as indicated in Figure 2.1 (Standard graphic symbols utilized throughout the electronics industry are shown in Appendix A.)

2.3 INFORMATION FLOW

Normally, a drafter first comes in contact with a block diagram in the form of a freehand sketch provided by an engineer or a designer, as illustrated in Figure 2.2. It is the drafter's responsibility to redraw the diagram with drafting tools and standard technical graphics

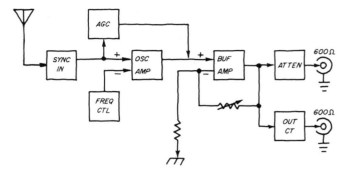

Figure 2.2
Freehand sketch of a block diagram.

practices so that the completed diagram meets the intended requirements.

The main flow of information, from "block to block," goes from left to right. In large or complex diagrams, the flow may take the form of several rows continuing from left to right, as shown in Figure 2.3. In all cases the direction of flow is identified by arrowheads within or at the end of each flow path line.

Flow path lines are drawn vertically and/or horizontally and are joined at the corners at a 90-degree angle. In addition, the path of the flow should be as direct as possible, avoiding the crossing of flow paths when possible. Figure 2.4 shows acceptable and unacceptable examples of flow paths. When required, auxiliary pieces of information may feed into the main flow of blocks. This auxiliary information may enter to or from any direction and be shown as either dotted or solid lines, as in Figure 2.5.

2.4 LINE CONVENTION

The most common line weight for block outlines as well as for flow paths is the solid, medium-weight line, as shown in Figure 2.6. If contrast or emphasis for some special reason is required, blocks only may be

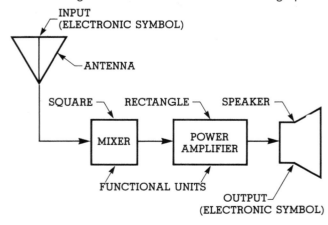

Figure 2.1
Graphic symbols for the block diagram.

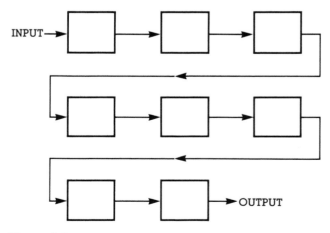

Figure 2.3
Information flow.

Figure 2.4
Acceptable and unacceptable flow paths.

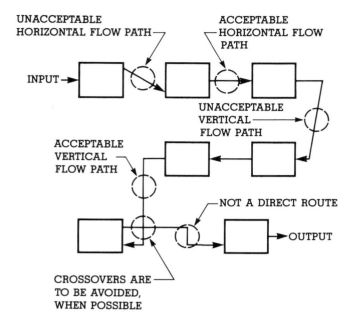

drawn with a heavy-weight line. Differences in recommended line weights are shown in Figure 2.7.

2.5 SIZE AND SHAPE OF BLOCKS

As indicated earlier, blocks are usually rectangular or square. There are two factors to consider in determining the size and shape of the block: (1) try to determine the number of blocks required to complete the diagram and select the correct size paper for the drawing; and (2) remember that the block needs to be large enough to accommodate its title or lettering.

When drawing blocks, you should consider drawing all rectangles or all squares; it is not good practice to mix squares and rectangles on the same diagram.

An extra-large block may be used, however, if several inputs and/or outputs are necessary to identify the functional flow, as illustrated in Figure 2.8. It is also common practice to draw rectangular blocks in a horizontal position as in Figure 2.9, with a ratio of length to height of blocks 1.5 to 1 or 2 to 1.

The most acceptable block diagram is one that has the same size blocks, equally spaced in vertical columns and horizontal rows, and shows clear, identifiable flow paths.

2.6 LETTERING

The block must contain the name of its components or function in the form of lettering. The lettering height

Figure 2.5
Auxiliary information flow.

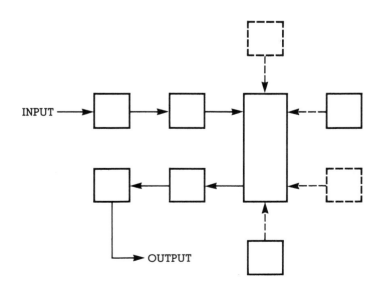

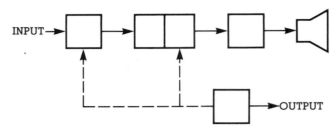

Figure 2.6
Commonly used line types.

depends largely on the final reduction size of the diagram. The words in each block should describe the function briefly and precisely. Words can be lettered on as many lines as necessary to accommodate all the needed information in each block. All block information may be spelled in full or abbreviated. Acceptable abbreviations for industries working under government contract are contained in American National Standards Institute (ANSI) Standard Y 1.1 or Military Standard (MIL-STD-12C). A list of the most common abbreviated forms appears in Appendix B. Figure 2.10 shows examples of acceptable forms of lettering.

Lettering should be centered both vertically and horizontally in each block, and you should not allow letters to contact or cross over the block outline. There should be sufficient space for each word. Never crowd the letters.

2.7 METHOD FOR DRAWING THE BLOCK DIAGRAM

This method represents standard practice as well as commonsense procedures for producing a block diagram.

1. Information flow is from left to right, with inputs on the left and outputs on the right of the drawing.

2. On complex diagrams, rows of blocks may continue to. as many rows as necessary to accommodate the diagram.
3. Most of the time, blocks are either rectangular or square.
4. Flow paths are drawn vertically and/or horizontally.
5. Flow paths should be as direct as possible.

2.7.1 DRAWING PROCEDURE

- Review all information provided for the preparation of the diagram (schematic or rough sketch of block diagram).
- Make a list of all the lettering and title information required for each block. Refer to American National Standards Institute (ANSI) Standard Y1.1 if abbreviations are required. Determine the height of lettering according to the final reduction size of the drawing.
- A preliminary layout should begin with the correct size drawing format to accommodate all the necessary blocks.
- Draw blocks with light-weight lines, or cut out paper blocks and move them around on the layout paper until you are satisfied with their location. Make certain that the blocks are large enough to accommodate the lettering.
- When you are satisfied with the block size and spacing, lightly draw in all the flow path lines.
- If the layout is acceptable, letter the titles or function of each block with freehand lettering or with a lettering guide.
- Complete the diagram by darkening all lines. Be sure to indicate the correct flow of information with arrowheads.

2.8 SUMMARY

The block diagram provides information to engineering and production personnel and may be used as an aid in

Figure 2.7
Contrasting line weights.

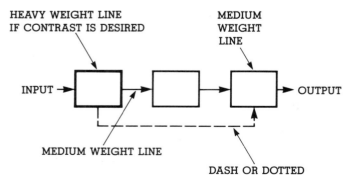

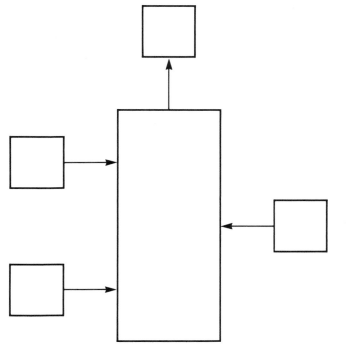

Figure 2.8
Mixed block sizes.

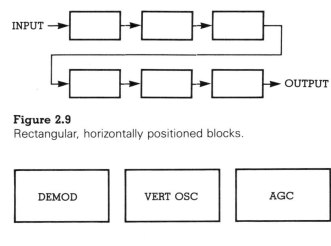

Figure 2.9
Rectangular, horizontally positioned blocks.

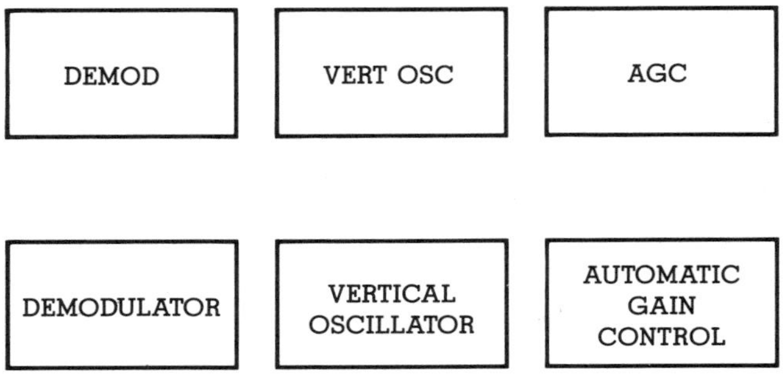

DEMOD	VERT OSC	AGC
DEMODULATOR	VERTICAL OSCILLATOR	AUTOMATIC GAIN CONTROL

Figure 2.10
Acceptable forms of lettering.

planning, designing, drafting, and fabricating electronic and electromechanical equipment. The basic unit in the diagram is the block, which provides component identification as well as its location in the system. Flow of information is provided in the form of flow path lines, and arrows indicate the direction of flow of information. We have identified acceptable forms of descriptive titles for use in blocks and a method for drawing the block diagram. Practice exercises for this section are on pages 185 through 195.

2.9 REVIEW EXERCISES

1. Describe the purpose of the block diagram:

2. Name and draw four (4) shapes that can be used when drawing block diagrams.

3. In what direction does the information flow on a block diagram?

4. In what direction are flow paths drawn?

5. What weight and type of line is used for flow path lines? Auxiliary lines?

6. How is block size determined?

7. What Standard is referred to when abbreviations are required for titles inside the blocks?

8. Identify the steps in the procedure for producing the block diagram.
a.

b.

c.

d.

e.

f.

g.

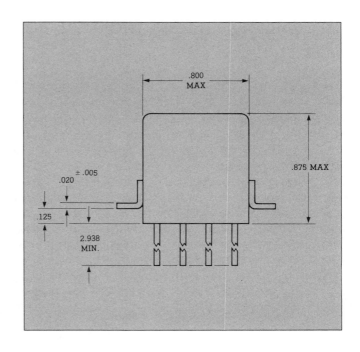

Section 3

THE CONTROL DRAWING

LEARNER OUTCOMES

The student will be able to:

- Name three types of control drawings

- List the differences in the three types of control drawings

- Draw control drawings from incomplete data

- Demonstrate learning through successful completion of practice and review exercises

3.1 PURPOSE AND FUNCTION

The control drawing outlines the configuration of a purchased part. It provides the overall mounting and mating dimensions as well as the shape of a component and usually contains enough specific data that a firm's purchasing department can readily procure the part. The control drawing is sometimes referred to as a "component drawing," "purchased part drawing," or an "outline drawing."

The control drawing is a necessary aid to the design engineer as well as to the drafter. The engineer relies on the electrical, environmental, and mechanical characteristics information provided by the control drawing and the drafter requires the physical dimensions so as to draw layouts. The control drawing is especially critical in electromechanical layouts and printed circuit board drawings, as we will see in Section 6, The Printed Circuit Board.

3.2 TYPES OF CONTROL DRAWINGS

In this section we will look at three types of control drawings: the envelope drawing, the specification control drawing, and the source control drawing. Similarities and differences will be identified.

3.2.1. THE ENVELOPE DRAWING

The envelope drawing illustrates a single component and specifies the configuration (shape) and performance envelope (parameters) without details of the part's internal construction. For all features other than those shown on the drawing, it is the responsibility of the manufacturer of the component to meet the required design specifications and performance data. "Design specifications" refers to the minimum performance requirements for satisfying the design of the final product or a system designed for the final product. "Performance data" is a group of physical as well as functional characteristics and environmental conditions that describe the operating conditions under which the component must perform. For example, conditions of temperature, altitude, speed, vibration, and shock are a few performance data characteristics that may be required for the component. Usually, the characteristics under which the part must perform are defined well enough so that several manufacturers can supply similar parts. Figure 3.1 depicts a typical envelope drawing and its associated required data. Please note that the words "ENVELOPE DRAWING" appear above the title block.

3.2.2. THE SPECIFICATION CONTROL DRAWING

This type of drawing identifies a commercially developed item for which all the engineering and test requirements can be met with items that are readily available through specialized industries. Included are manufacturers that produce typical components such as potentiometers, switches, special controls and motors, hydraulic and pneumatic devices, hermetically sealed and potted components, and relays. Figure 3.2 shows a specification control drawing and illustrates the kind of information required to purchase a part called a relay. Besides the notation that identifies the drawing as a SPECIFICATION CONTROL DRAWING, there is a list headed SUGGESTED SOURCES OF SUPPLY that identifies each manufacturer who has been accepted as having met all the drawing's requirements. Other areas that apply in this example are the component description, electrical requirements, mechanical requirements, document requirements (military specifications), marking requirements, envelope dimensions, mounting dimensions, and a circuit diagram.

3.2.3 THE SOURCE CONTROL DRAWING

The information on the source control drawing restricts procurement to a specific source or sources. Engineering requirements are controlled by the restrictions imposed by the specific design group or activity. The purchase of the component is approved as a result of the test or evaluation that has proven satisfactory for the application. This type of drawing is used when it is necessary to limit the source of supply for a part for use in a critical application. If an item can be completely defined for all applications, then a specification control drawing is appropriate. If not, then a source control drawing will be required. Figure 3.3 identifies a source control drawing for a transformer. The information on the drawing includes the name and address of the manufacturer as an approved source of supply, outline and mounting dimensions for the transformer, specifics about the transformer (such as descriptive, electrical construction, environmental and marking requirements), the notation SOURCE CONTROL DRAWING immediately above the title block, and a note that states "ONLY THE ITEM DESCRIBED ON THIS DRAWING WHEN PROCURED FROM THE MANUFACTURER(S) LISTED HEREON IS APPROVED BY (name of company or design activity giving approval) FOR USE IN THE APPLICATION(S) SPECIFIED HEREON. A SUBSTITUTE ITEM SHALL NOT BE USED WITHOUT PRIOR TESTING AND APPROVAL BY

NOTES:
1. DESCRIPTION: A TYPE 2N1854 PNP TRANSISTOR
   WITH CONTROLLED DIAMETER WELDABLE WIRE LEADS.

2. REQUIRED DOCUMENTS: THE FOLLOWING DOCUMENTS
   OF THE ISSUE IN EFFECT ON DATE OF INVITATION FOR
   BIDS, FORM A PART OF THIS DRAWING TO THE
   EXTENT SPECIFIED HEREIN:
       MIL-S-19491
       MIL-S-19500/172
       QQ-N-281
       MIL-STD-130

3. REQUIREMENTS: THE -001 SHALL MEET THE
   REQUIREMENTS OF MIL-S-19500/172 EXCEPT:
   3:1 LEAD DIAMETER SHALL BE PER FIGURE 1.
   3.2 LEAD MATERIAL SHALL BE A NICKEL-COPPER
       ALLOY PER QQ-N-281.
   3.3 MARKING FOR IDENTIFICATION SHALL BE
       IN ACCORDANCE WITH MIL-STD-130.

4. PREPARATION FOR DELIVERY SHALL BE IN
   ACCORDANCE WITH MIL-S-19491, LEVEL C.

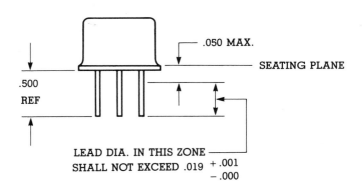

ENVELOPE DRAWING

DRAWN		TRANSISTOR, SWITCHING, GERMANIUM, PNP	
CHECKED			
ENGINEER			
DATE	APPROVED		
CONTRACT NO.		SIZE **A**	DRAWING NO. N100200
ENG. RELEASE			
		SCALE FULL	SHEET 1 OF 1

Figure 3.1
Envelope drawing.

NOTES:
1. DESCRIPTION: AN UNSEALED 400 AMP SOLENOID OPERATED
   SPST NORMALLY OPEN CLASS AS RELAY.

2. REQUIRED DOCUMENTS: THE FOLLOWING DOCUMENT(S) OF THE
   ISSUE IN EFFECT ON DATE OF INVITATION FOR BIDS, FORM A PART
   OF THIS DRAWING TO THE EXTENT SPECIFIED HEREIN.
   MIL-R-00045

3. REQUIREMENTS:
   3.1 ELECTRICAL
      3.1.1 NOMINAL OPERATING VOLTAGE
         COIL ....24 -28 VOLTS DC
         CONTACTS ....7 VOLTS DC CAPABLE OF WITHSTANDING
                                  90 VOLTS DC AT NO LOAD
      3.1.2 MAX COIL VOLTAGE ....31 VOLTS DC FOR 30 MINUTES
      3.1.3 MAX COIL CURRENT ....1.6 AMPS AT 31 VOLTS DC AND 77° F
      3.1.4 RATED CONTACT LOAD .....400 AMPS, CAPABLE OF OPENING AT
                                   A MOMENTARY LOAD OF 600 AMPS
   3.2 MECHANICAL
      3.2.1 DIMENSIONAL REQUIREMENTS PER FIGURE 1
      3.2.2 MAXIMUM WEIGHT -2.5 LB.
      3.2.3 REQUIREMENTS OF MIL-R-00045 APPLY

3.3 MARKING: PERMANENTLY AND LEGIBLY MARK
    PART WITH MANUFACTURER'S PART NUMBER

SUGGESTED SOURCES OF SUPPLY

VENDOR
K.C. JONES INC. 123 A ST. KANSAS CITY, MO CODE IDENT.
EUCLID ELEC CO. 624 CLAY ST. BUFFALO, N.Y. CODE IDENT.

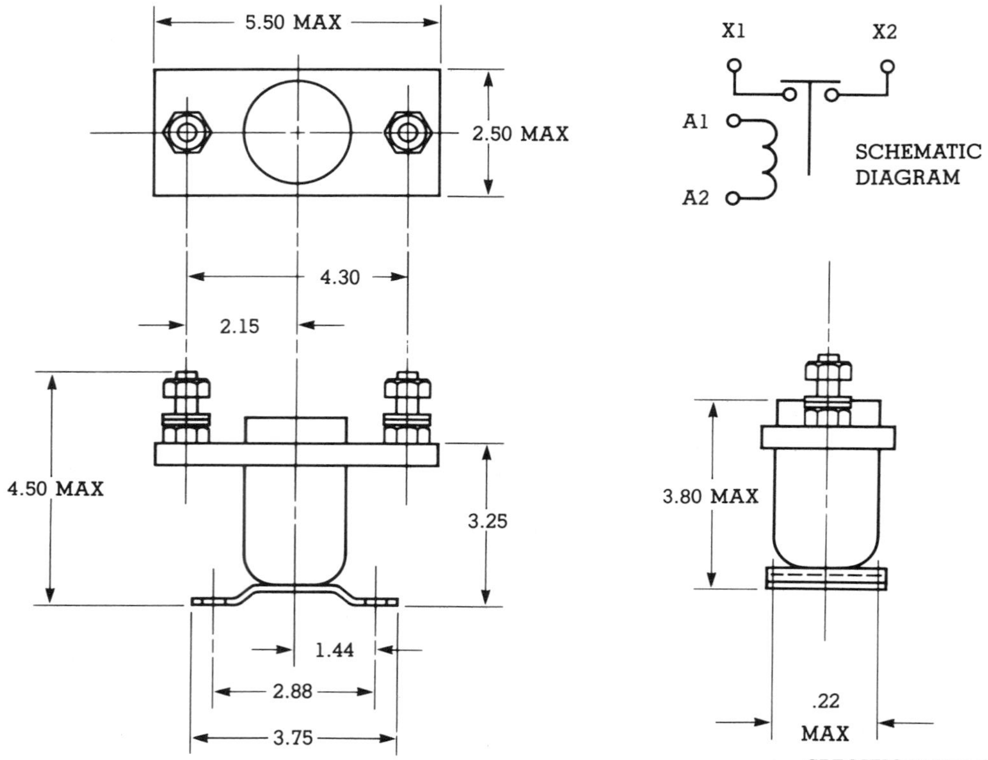

5.50 MAX

2.50 MAX

4.30

2.15

X1   X2

A1

A2

SCHEMATIC
DIAGRAM

4.50 MAX

3.25

1.44

2.88

3.75

3.80 MAX

.22
MAX

SPECIFICATION CONTROL DWG

DRAWN		RELAY, SOLENOID
CHECKED		
ENGINEER		
DATE	APPROVED	
CONTRACT NO.		SIZE	DRAWING NO.
		A	N100201
ENG. RELEASE			
		SCALE	SHEET 1 OF 1

Figure 3.2
Specification control drawing.

NOTES:
1. DESCRIPTION: A GENERAL PURPOSE SEALED TRANSFORMER
   MEETING THE REQUIREMENTS OF MIL-T-27 FOR GRADE 4,
   CLASS R, LIFE EXPECTANCY X.

2. REQUIRED DOCUMENTS: THE FOLLOWING DOCUMENTS OF
   THE ISSUE IN EFFECT ON DATE OF INVITATION FOR BIDS, FORM
   A PART OF THIS DRAWING TO THE EXTENT SPECIFIED HEREIN.
   MIL-T-27
   MIL-P-8585
   MIL-E-15090
   MIL-T-00073
   FED-STD-595

3. REQUIREMENTS
   3.1 ELECTRICAL
       3.1.1 PRIMARY: 109, 115, AND 121 VOLTS RMS, 57 TO 63 HERTZ
             SINGLE PHASE
       3.1.2 SECONDARY: CENTERTAPPED AT TERMINAL 6. VOLTAGE,
             250 VOLTS RMS LOADED, DIRECT CURRENT RESISTANCE,
             2.0 OHMS MAX. LOAD CURRENT, 1.7 AMPERES RMS.
       3.1.3 AN ELECTROSTATIC SHALL BE PROVIDED BETWEEN THE
             PRIMARY AND SECONDARY. SHIELD, CORE, AND CORE
             BRACKETS SHALL BE ELECTRICALLY AND MECHANICALLY
             SECURED TO MOUNTING DEVICES
   3.2 CONSTRUCTION
       3.2.1 HERMETICALLY SEALED METAL CASE
       3.2.2 FINISH PRIME WITH ONE COAT ZINC CHROMATE, SPEC
             MIL-P-8585 AND FINISH WITH ONE COAT GREY ENAMEL,
             SPEC MIL-E-15090. COLOR PER FED-STD-595 COLOR CHIP
             16376
   3.3 ENVIRONMENTAL
       3.3.1 AMBIENT TEMPERATURE RANGE, -10 TO +65° C.
       3.3.2 MAXIMUM OPERATING TEMPERATURE 105°C.
   3.4 MARKING: PERMANENTLY AND LEGIBLY MARK THE ITEM
             WITH TERMINAL NUMBERS SHOWN AND THE ELECTRONICS
             DIVISION PAR NUMBER, V00078 -001, AND CODE
             IDENTIFICATION NUMBER 58189.

4. APPROVED SOURCE OF SUPPLY: ONLY THE ITEM DESCRIBED
   ON THIS DRAWING WHEN PROCURED FROM THE VENDOR(S) LISTED
   HEREON IN APPROVED FOR USE IN THE APPLICATION(S) SPECIFIED
   HEREON, A SUBSTITUTE ITEM SHALL NOT BE USED WITHOUT PRIOR
   TESTING AND APPROVAL BY THE PROCURING ACTIVITY.

DASH NO.	VENDOR	VENDOR'S ITEM NO.	APPLICATION
-001	TRIAX CO. 690 MARKET ST. CHICAGO, ILL	6041H7	AMPLIFIER 553-MOZ

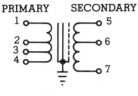

PRIMARY   SECONDARY

SCHEMATIC
DIAGRAM

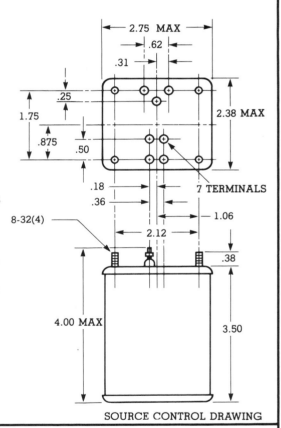

SOURCE CONTROL DRAWING

DRAWN		
CHECKED		
ENGINEER		
DATE	APPROVED	

TRANSFORMER, SHIELDED

CONTRACT NO.		SIZE	DRAWING NO.
ENG. RELEASE		A	N100202
		SCALE	SHEET 1 OF 1

Figure 3.3
Source control drawing.

(name of company or design activity giving approval) OR BY (name of the Government procuring activity)." Otherwise, the requirements are the same as those shown on the specification control drawing. *Note:* You should not use a source control drawing if a specification control drawing will meet the design needs.

3.3 DRAWING PREPARATION

Control drawings are prepared as a result of a particular need on an engineering project. The required type of control drawing depends on whether the desired part is a commercial component or one that will be manufactured under strict military specifications. In either case, the drafter should receive the appropriate preliminary information from the project leader, drafting supervisor, or the project's lead designer before beginning the drawing. The information may take the form of a drawing from a manufacturer's catalog or a sketch that includes the component's various performance and physical characteristics.

Control drawings are produced on "A" size (8½ × 11) drawing material. Because environmental, electrical, mechanical, marking, and quality control information may be required on the drawing, you may need several sheets, as the specification control drawing in Figure 3.4 (pp. 37–40) shows.

3.4 SUMMARY

The three types of control drawings discussed in this section vary depending on the application of the component they represent, the required design specifications or performance data, and whether the component is commercially available or developed for a specific, controlled project. A control drawing may require a single page or several sheets depending upon the necessary specifications.

Practice exercises for this section are on pages 197 through 205.

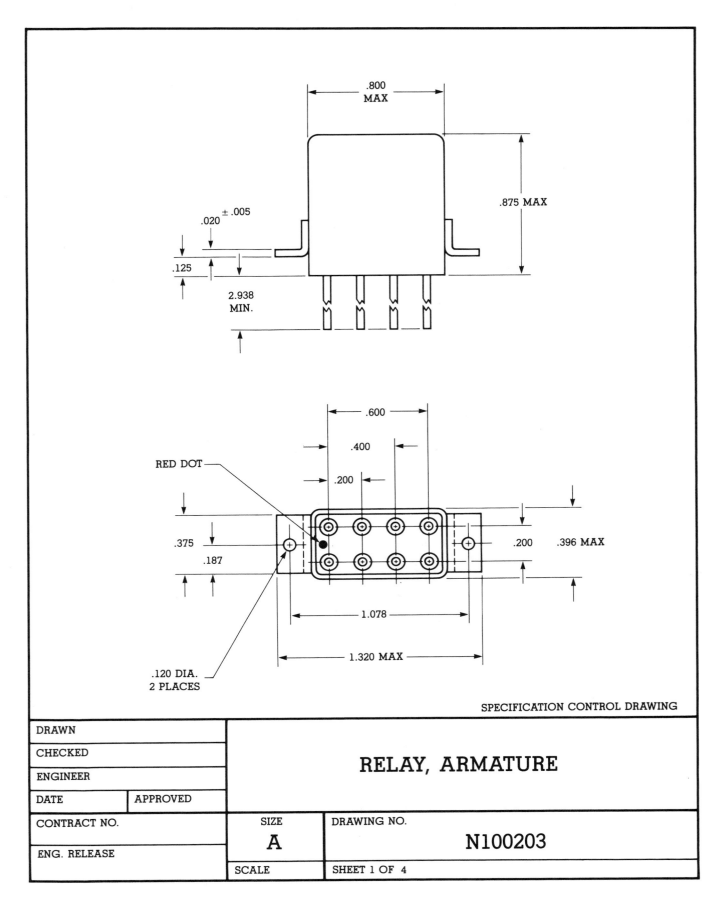

.800
MAX

.875 MAX

± .005
.020

.125

2.938
MIN.

RED DOT

.600

.400

.200

.375

.187

.200   .396 MAX

1.078

1.320 MAX

.120 DIA.
2 PLACES

SPECIFICATION CONTROL DRAWING

DRAWN			
CHECKED		**RELAY, ARMATURE**	
ENGINEER			
DATE	APPROVED		
CONTRACT NO.		SIZE **A**	DRAWING NO. **N100203**
ENG. RELEASE		SCALE	SHEET 1 OF 4

Figure 3.4 α
Multi-sheet control drawing.

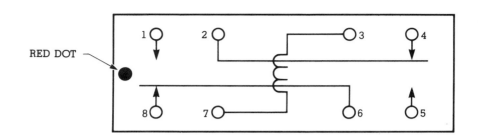

1. DESCRIPTION :
   RELAY, ARMATURE, HERMETICALLY SEALED, IN ACCORDANCE WITH THE
   REQUIREMENTS OF TYPE RY4—483L PER MIL-R-5757D EXCEPT AS NOTED.

2. ELECTRICAL REQUIREMENTS:
   2.1 CONTACTS:
      2.1.1 ARRANGEMENT:
         2 FORM C (DPDT) (SHOWN IN THE
         DE-ENERGIZED CONDITION, SEE
         FIGURE 1).
      2.1.2 RATINGS
         2.1.2.1 RATED LOADS
            2.1.2.1.1 28VDC:
               3A, RESISTIVE; 1A INDUCTIVE
               (100 MILLIHENRYS).
            2.1.2.1.2 115VRMS: 60 CPS:
               1A RESISTIVE; 0.5A INDUCTIVE (60%
               P.F.) CASE INSULATED FROM GROUND.
         2.1.2.2 LOW LEVEL LOAD:
            1 MILLIVOLT OPEN CIRCUIT,
            1 MICROAMPERE CLOSED CIRCUIT.
      2.1.3 LIFE:
         100,000 OPERATIONS, MINIMUM FOR ANY OF THE ABOVE LOAD
         CONDITIONS WITH NOMINAL COIL INPUT AT MAXIMUM RATED AMBIENT
         TEMPERATURE. ONE OR MORE MISSES SHALL BE CONSIDERED A
         FAILURE. A MISS IS DEFINED AS CONTACT RESISTANCE EXCEEDING
         THAT SPECIFIED IN PARAGRAPHS 2.1.4.1.2 OR 2.1.4.2 AS
         APPLICABLE.

CONTRACT NO.		SIZE	DRAWING NO.	
		A		N100203
ENG. RELEASE				
		SCALE	SHEET 2 OF 4	

Figure 3.4 b
Multi-sheet control drawing.

2.1.4 RESISTANCE: (MAXIMUM)

    2.1.4.1 RATED LOADS: (TO BE DETERMINED AT FULL RATED LOAD AT 6 VDC).

        2.1.4.1.1 INITIAL: 0.050 OHMS
        2.1.4.1.2 DURING LIFE AND FOLLOWING
    LIFE TEST: 0.100 OHMS

        2.1.4.2 LOW LEVEL LOAD: 100 OHMS (TO BE DETERMINED AT 1 UA CLOSED CIRCUIT AT 1 MV OPEN CIRCUIT.

2.2 COIL INPUT:

2.2.1	MAXIMUM PICKUP:	9.25 VDC
2.2.2	NOMINAL:	12.6 VDC
2.2.3	MAXIMUM CONTINUOUS:	17 VDC AT +25°C
2.2.4	MINIMUM DROP-OUT:	0.68 VDC AT +25°C
2.2.5	COIL RESISTANCE:	200 OHMS, ±10% AT 25°C

2.3 TIMING (MAXIMUM CONTACT TRANSFER TIME, NOT INCLUDING BOUNCE DURATION WITH NOMINAL COIL INPUT AT 25°C).

2.3.1	OPERATE:	4.5 MILLISECONDS
2.3.2	RELEASE:	2 MILLISECONDS

2.4 CONTACT BOUNCE (MAXIMUM DURATION AT 25°C).

2.4.1	OPERATE:	2.5 MILLISECONDS
2.4.2	RELEASE:	4 MILLISECONDS

3. MECHANICAL REQUIREMENTS:

3.1 RELAY SHALL CONFORM TO THE DIMENSIONAL REQUIREMENTS.

3.2 ENCLOSURE SHALL BE HERMETICALLY SEALED.

3.3 TERMINAL PULL TEST SHALL BE 5 ±.5 POUNDS.

3.4 DEGASSING IS REQUIRED.

3.5 WEIGHT: 0.62 OZ. MAXIMUM

4. ENVIRONMENTAL REQUIREMENTS:

4.1 RELAYS SHALL MEET THE ENVIRONMENTAL REQUIREMENTS OF MIL-R-5757D FOR THE TYPE DESIGNATIONS AS NOTED.

CONTRACT NO.	SIZE	DRAWING NO.	
	A		**N100203**
ENG. RELEASE			
	SCALE	SHEET 3 OF 4	

Figure 3.4 c
Multi-sheet control drawing.

4.1.1	TEMPERATURE CLASS B:			-65°C TO +125°C
4.1.2	SHOCK:			GRADE - 3 (50G'S)
	4.1.2.1	CHATTER:		10 USEC MAXIMUM
4.1.3	VIBRATION CHARACTERISTIC:			4 (10 TO 2000 CPS)
	4.1.3.1	CHATTER:		10 USEC MAXIMUM

5. <u>MARKING</u>:
RELAYS SHALL BE PERMANENTLY AND LEGIBLY MARKED IN ACCORDANCE WITH
MIL-R-5757D AND CONTAIN THE FOLLOWING INFORMATION:

5.1 MILITARY TYPE DESIGNATION WHEN APPLICABLE

5.2 RATED (NOMINAL) COIL VOLTAGE

5.3 COIL RESISTANCE

5.4 CONTACT RATING(S)

5.5 CIRCUIT DIAGRAM, WITH TERMINAL IDENTIFICATION BY DESIGNATION AND/OR
PHYSICAL ARRANGEMENT.

5.6 VENDOR NAME OR SYMBOL

5.7 VENDOR PART NUMBER

5.8 DATE (CODE) OF MANUFACTURE

6. <u>APPROVAL</u>:
APPROVAL IS REQUIRED BY THE ENGINEERING DEPARTMENT.

7. <u>QUALITY ASSURANCE PROVISIONS</u>:
QUALITY ASSURANCE SHALL BE IN ACCORDANCE WITH MIL-R-5757D

CONTRACT NO.	SIZE	DRAWING NO.	
	A	N100203	
ENG. RELEASE			
	SCALE	SHEET 4 OF 4	

Figure 3.4 d
Multi-sheet control drawing.

3.5 REVIEW EXERCISES

1. What purpose does the control drawing serve in an engineering department?

2. Name three types of control drawings.

3. What does the term "design specifications" mean?

4. Define "performance data." Give four examples of performance data.

5. What type of control drawing identifies a commercial or a manufacturer-developed item? Give five examples of commercially available items.

6. Identify six types of requirements shown on a typical specification control drawing.

7. What type of component drawing is needed if a component is approved as a result of tests or evaluation for a stated critical application?

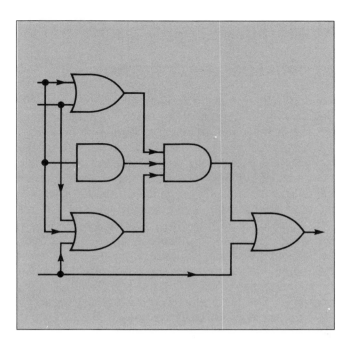

Section 4

THE LOGIC DIAGRAM

LEARNER OUTCOMES

The student will be able to:

- List and describe several logic functions

- Draw several logic symbols using a drafting template

- Draw several logic diagrams using drafting tools and equipment

- Describe four types of logic diagrams

- Demonstrate learning through successful completion of practice and review exercises

4.1 PURPOSE AND FUNCTION

The logic circuit may be considered a decision-making device that is capable of recognizing specific input level conditions and providing an output in the form of a voltage level or pulse. The logic diagram illustrates, through flow sequences, the operation of logic circuits by means of symbols representing logical functions. It identifies the details of signal flow and control that exist in a system of two-state devices.

A logic diagram is drawn with a few basic logic symbols. These symbols, representing such elements as gates and flip-flops, are used again and again in various combinations to perform different functions. Understanding the meaning of these basic symbols helps one determine the operation of complex systems.

Logic diagrams serve two basic needs. They (1) provide a source of information for rapidly localizing a malfunction to a single unit within a system; and (2) are excellent tools for training personnel in the operation and maintenance of electromechanical equipment.

4.2 TYPES OF LOGIC DIAGRAMS

Industry uses several types of logic diagrams.

4.2.1. THE BASIC LOGIC DIAGRAM

The basic logic diagram identifies logic functions with no reference to physical considerations. This diagram consists of logic symbols that show all logic relationships as simply as possible. Nonlogic functions are not shown. (We define nonlogic functions as those devices, such as resistors and capacitors, that are not represented by a standard logic symbol.)

4.2.2 THE DETAILED LOGIC DIAGRAM

The detailed logic diagram depicts all logic as well as nonlogic functions, socket locations, pin numbers, test points, and other physical elements necessary to describe a system's electrical and physical aspects.

4.2.3 THE RECTANGULAR SHAPE LOGIC DIAGRAM

This diagram is a detailed or basic logic diagram in which both logic and nonlogic functions are identified by the use of rectangles.

4.2.4 THE DISTINCTIVE SHAPE LOGIC DIAGRAM

This diagram is a detailed or basic logic diagram in which outlined enclosures with distinctive shapes rep-resent both logic and nonlogic functions. This type of diagram has recently become popular because the distinctive shapes simplify the reading of complex diagrams.

4.3 LOGIC STATES

In describing logic states, it is important to understand that some terms are used interchangeably. Normally, there are only two "states" or "conditions" in the logic system, and they may be referred to in several ways. They include:

HI	or	LO
1	or	0 (zero)
ON	or	OFF
significant	or	reference

where HI signals are frequently referred to as the 1 state or the ON state, while LO signals are often referred to as the 0 (zero) or the OFF state. The 1 state is also often called the significant state, while the 0 (zero) state is referred to as the reference state. The exception to the two-state rule is shown in Table 4.1 and described as a *tri-state buffer*. This logic device is used extensively in microprocessor systems.

The binary numbers 1 and 0 (zero) are referred to as *bits* and may be defined as:

- An abbreviation of a binary digit
- A single character of a language employing only two distinct kinds of characters
- A unit of storage capacity

For convenience, we will refer only to the HI and LO states in this worktext.

Logic techniques are used to perform many functions. The function and operation of logic circuits are described with symbols and special terminology. Table 4.1 helps to explain these symbols and terms. The symbols in the table are the most common ones identified in military standard MIL-STD-806B.

4.4 SYMBOL PRESENTATION TECHNIQUES

In a logic diagram, it is important to identify each function correctly. Therefore, consideration should be given to the following points before beginning the diagram.

- The orientation of a symbol does not alter its meaning.
- The line weight of a symbol does not affect the meaning of the symbol. Logic diagrams are not drawn to a specific scale. Symbols are drawn with the template in Figure 4.1.

Table 4.1
Basic logic reference.

Function/Device	Symbol	Description	Input/Output Table
AND gate	A, B → X	A device with two or more inputs and one output. The output of the AND gate is HI if and only if all of the inputs are at the HI state.	**Input / Output** A · B · X LO · LO · LO LO · HI · LO HI · LO · LO HI · HI · HI
NAND gate	A, B → X	A device with two or more inputs and one output. The output of the NAND gate is LO if and only if all of the inputs are at the HI state.	**Input / Output** A · B · X LO · LO · HI LO · HI · HI HI · LO · HI HI · HI · LO
OR gate	A, B → X	A device with two or more inputs and one output. The output of the OR gate is HI if one or more of the inputs are at the HI state.	**Input / Output** A · B · X LO · LO · LO LO · HI · HI HI · LO · HI HI · HI · HI
NOR gate	A, B → X	A device with two or more inputs and one output. The output of the NOR gate is LO if one or more of the inputs are at the HI state.	**Input / Output** A · B · X LO · LO · HI LO · HI · LO HI · LO · LO HI · HI · LO
EXCLUSIVE OR	A, B → X	A device which produces a signal only when one of its input signals occur.	**Input / Output** A · B · X LO · LO · LO LO · HI · HI HI · LO · HI HI · HI · LO
FLIP-FLOP	R → O, S → 1	A device which stores a single bit of information. It normally has two inputs set (S) and reset (R) and two outputs (O) and (I).	
AMPLIFIER	A → X	A device with one input and one output. If output is HI input is HI. The symbol represents a linear or non-linear current.	**Input / Output** A · X LO · LO HI · HI
INVERTER	A → X	A device with one input and one output. The input state is always opposite to the output state. It is an inverted amplifier.	**Input / Output** A · X LO · HI HI · LO
TRI-STATE BUFFER	A → X, $\overline{CE}$	A device that has three states. The output is HI when the enable input is HI and LO when the enable input is LO. The device operates in its third state when the enable input is not active.	**Input / Output** A · CE · X LO · LO · HI HI · LO · LO LO · HI · HI Z HI · HI · HI Z

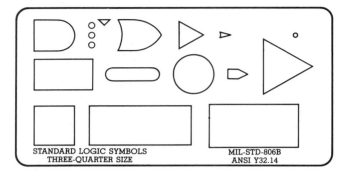

Figure 4.1
Logic symbols template.

■ A symbol may be drawn to any proportional size that satisfies the drawing, depending upon its ultimate reduction or enlargement requirement. The American National Standards Institute (ANSI) recommends several sizes of templates, ranging from full scale to one-quarter scale, for drawing the logic diagram. Their reference standard is ANSI Y32.14, the same as that called for in military standard MIL-STD-806B. An example of how the template is used is demonstrated in Figure 4.2.

4.5 TAGGING LINES

On a logic diagram, logic symbols are identified by tagging or identification lines located within the symbol, as indicated in Figure 4.3. The tagging lines identify the logic function and its location within the equipment. In Figure 4.3, "AND" represents the type of logic function, "3" identifies the drawing sheet se-

Figure 4.2
How to use a logic symbols template.

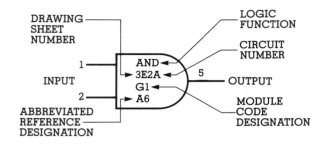

Figure 4.3
Tagging lines.

quence number, "E2A" is the identifying number given to the circuit, "G1" indicates the module code designation, and "A6" is an abbreviated reference designation for the assembly, in this case, assembly #6 in the system. In addition, pin numbers for both input and output are placed around the exterior of the symbol.

4.6 FUNCTION IDENTIFICATION LETTER COMBINATIONS

The following letter combinations are specified in MIL-STD-806B and are used to identify functions on logic diagrams.

Function	Letter Combination
Blocking Oscillator	BO
Clear	C
Cathode Follower	CF
Emitter Follower	EF
Flip-Flop	FF
Number of Bits or Stages	(N)
Register with N Stages	RG(N)
Set	S
Shift Register with N Stages	SR(N)
Single-Shot	SS
Schmitt Trigger	ST
Trigger or Toggle	T

The appropriate letter combination is placed within the symbol as shown in Figure 4.4. The example identifies a Flip-Flop.

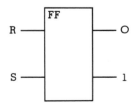

Figure 4.4
Function identification.

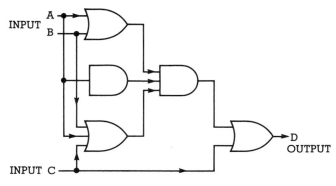

Figure 4.5
Signal flow.

4.7 SIGNAL FLOW

The signal flow should be direct and the connecting paths drawn as short as possible. Primary inputs are placed on the left side of the drawing and primary outputs on the right, as illustrated in Figure 4.5. The direction of signal flow is indicated by arrowheads placed on the signal path lines, as indicated in Figure 4.6. The arrowheads should not touch the input or the output of the graphic symbol outline. This point is illustrated in Figure 4.7.

Signal paths that cross should be shown as in Figure 4.8, and the junction of signal paths should be

Figure 4.6
Arrowhead placement.

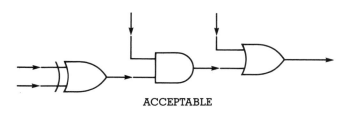

ACCEPTABLE

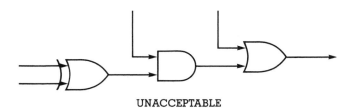

UNACCEPTABLE

Figure 4.7
Acceptable and unacceptable placement of arrowheads.

drawn as in Figure 4.9. The junction dot should be solid and approximately $\frac{3}{32}$ inches in diameter.

4.8 METHOD FOR DRAWING THE LOGIC DIAGRAM

Usually the engineer, the chief drafter, or the lead designer furnishes preliminary data to the drafter in the form of a freehand sketch of a logic diagram. The drafter should:

- Carefully review the information provided to make certain that the diagram is clear, symbols legible, and information complete.
- Examine the sketch to see if the diagram can be simplified, perhaps by minimizing crossovers, providing more appropriate line path location, and increasing spacing between lines.
- Select the correct size and type of drawing paper according to the complexity of the project, the size of symbols to be used, and the final reduction size of the drawing, if known.
- Start the diagram by laying out groupings or major areas and symbols indicated by the sketch, using light-weight lines.
- Draw connecting vertical and horizontal flow lines to all required symbols and terminals.
- Examine the diagram at this point to see if it compares favorably with the information provided.
- If satisfied, you can "heavy-up" or darken all lines to the appropriate line weight. Add necessary lettering, either freehand or with a lettering guide, to complete the logic diagram.

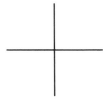

Figure 4.8
Crossing signal path.

Figure 4.9
Signal path junction.

4.9 SUMMARY

The logic diagram depicts the operation of logic circuits by means of symbols that represent logical functions. Only a few basic symbols are required to identify these logic functions. Several types of logic diagrams and techniques for identifying each function were described. We have also discussed a detailed method for drawing the logic diagram.

Practice exercises for this section are on pages 207 through 215.

4.10 REVIEW EXERCISES

1.   What is the purpose of the logic diagram?

2.   Name and describe four types of logic diagrams.

3.   What is the advantage of using distinctive shapes for functions on a logic diagram?

4.   What is the purpose of the binary numbers 0 (zero) and 1 (one) in a logic diagram?

5.   Draw and identify 4 different logic symbols.

6.   The following are either true (T) or false (F) statements. Circle T or F for the correct answer.

   a.   The orientation of a logic function symbol does alter its meaning.   T   F

   b.   The weight of line of a logic function symbol does not affect the meaning of the symbol.   T   F

   c.   Logic function templates are available in several different scales.   T   F

7. Define FLIP-FLOP, AND gate, NOR gate, and AMPLIFER logic symbols and their functions.

8. What is the purpose of the tagging line?

9. What are the correct letter combinations for the following logic functions?

Blocking Oscillator   _____

Single-Shot   _____

Shift Register with N Stages   _____

Emitter Follower   _____

Flip-Flop   _____

Schmitt Trigger   _____

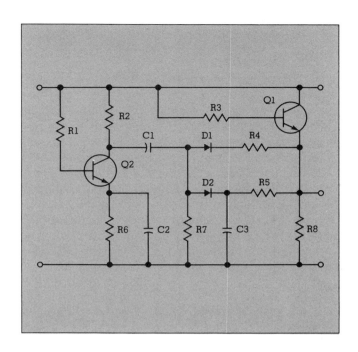

Section 5

THE SCHEMATIC DIAGRAM

LEARNER OUTCOMES

The student will be able to:

- Identify and draw electronics graphic symbols using a symbol template

- Identify component reference designations

- Identify component value information and component sequence numbers

- Draw partial and complete schematic diagrams utilizing appropriate layout methods

- Demonstrate learning through successful completion of practice and review exercises

5.1 PURPOSE AND FUNCTION

The schematic diagram graphically describes, through a series of lines and forms, those circuits that may contain electrical, electronic, or electromechanical devices or components identified by standard graphic symbols. These symbols are placed so as to allow the circuit to be traced in a functional sequence without regard to the physical location, shape, or size of the items or components. The schematic diagram should normally be arranged in a drawing format so that its completed form will read functionally from left to right and top to bottom. Large and/or complex circuitry may require a number of layers, sheets, or layouts. Each subsequent layer should read from left to right for convenience and appropriate flow of information. The overall diagram, therefore, will read from the left (inputs) to the right (outputs) in its completed form. An example is shown in Figure 5.1.

The overall goal is to produce a drawing called a *schematic diagram* resulting from a circuit layout that follows the transmission path from input to output or in the order of functional sequence and may be part of an electrical/electronic unit, subsystem, or system.

5.2 GRAPHIC SYMBOLS

Electrical and electronic symbols are used in single-line diagrams, schematic diagrams, logic diagrams, logic-schematic diagrams, and, at times, on connection or wiring diagrams. Each diagram graphically represents the functioning of electrical, electronic, or electromechanical components of a specific circuit. To conform to government or military contracts, one should use the template illustrated in Figure 5.2, which complies with American National Standard Institute (ANSI) Y32.2, Graphic Symbols for Electrical and Electronics Diagrams.

The use of graphic symbol templates is recommended for speed and uniformity in the drafting process. The use of templates is a basic skill one must

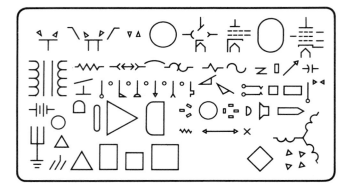

FIGURE 5.2
Graphic symbols template—ANSI Y32.2.

master to produce a proper and well drawn schematic. The many templates that can be used in drafting work are available in drafting and graphic art supply stores and stationery and book stores. Graphic symbols commonly used throughout the electronics industry are shown in Appendix A.

5.3 CONDUCTOR PATHS

5.3.1 LINE CONVENTION

Selection of line widths and lettering heights is determined by the legibility requirements of the final reduced size of the schematic diagram. A medium-weight line is recommended for general use on schematic diagrams. A medium line should also be used for mechanical linkages and shielding.

To emphasize special features such as mechanical grouping or main transmission lines, a thick line is required to provide the desired contrast. Bus bar or cable routing is identified by an extra-thick line. It is important to maintain consistent line weight throughout the entire schematic. Figure 5.3 gives examples of line convention for schematic use.

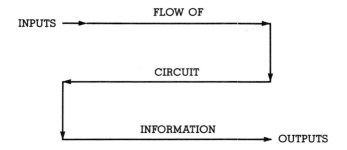

FIGURE 5.1
Circuit information flow.

LINE APPLICATION	LINE WIDTH
FOR GENERAL USE AS CONDUCTOR PATH	MEDIUM
BUS BAR, CABLE ROUTING	EXTRA THICK
MECHANICAL CONNECTION, SHIELDING, AND FUTURE CIRCUITS	MEDIUM
MECHANICAL GROUPING AND FIGURE ENCLOSURES	THICK

FIGURE 5.3
Line convention on schematics.

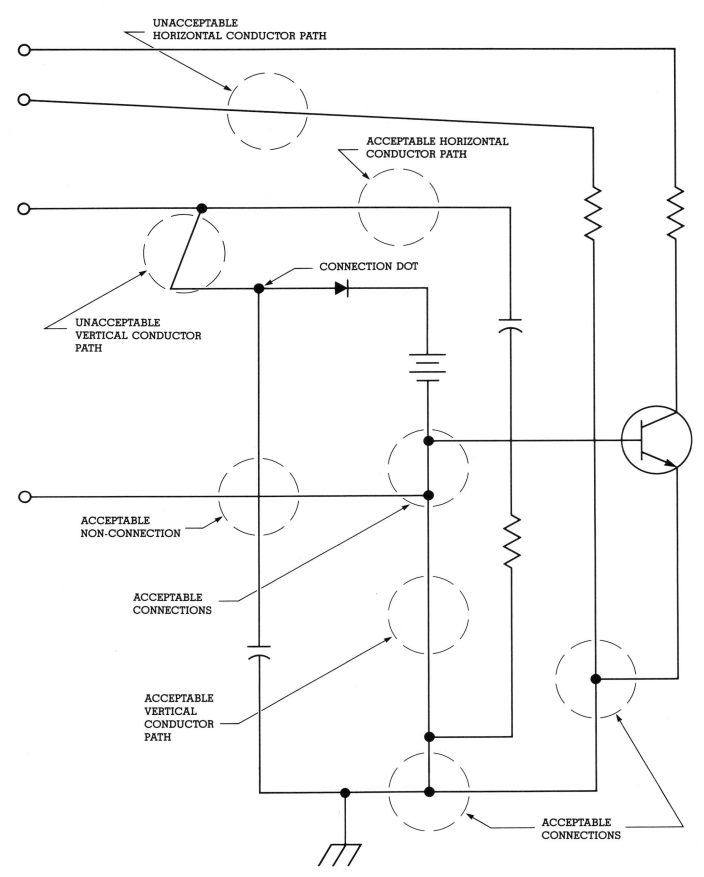

FIGURE 5.4
Acceptable and unacceptable conductor paths, connections, and nonconnections.

53

5.3.2   CONNECTING LINES

Connecting lines, also known as conductor paths, should be drawn only horizontally and vertically. Care should be taken to keep crossover paths and bends to a minimum. In addition, you should avoid connecting more than four lines at any one point unless there is a severe space problem. Acceptable and unacceptable methods for connections and nonconnections are shown in Figure 5.4. Connection dots should be a solid circle with a diameter of approximately ³⁄₃₂ inches.

5.4   REFERENCE DESIGNATIONS

All graphic symbols of separately replaceable parts should be identified by their appropriate reference designation. The designation provides alphanumeric identifiers. The basic reference information for a component such as a resistor, capacitor, or transistor consists of a class designation letter followed by a component sequence number. Recommended placement of reference designations and component values, when required, may be on the left or right, above or below the graphic symbol, as shown in Figure 5.5. A list of class designation letters appears in Appendix C.

The placement or direction of a graphic symbol on a drawing does not alter its meaning. Symbols that lend themselves to rotation can be so arranged for simplification of the circuit. The size of a symbol does not affect its meaning. Symbols may be drawn to any proportionate size compatible with the drawing and reproduction sizes. All efforts should be made, however, to assure that the same size symbols are used throughout the entire schematic drawing.

The flow of component sequence numbers of reference designations is standard throughout the electronics industry. The flow should begin with the lowest component number in the upper left-hand area of the schematic drawing and continue consecutively from left to right throughout the diagram, following the functional sequence of the schematic as indicated in Figure 5.6. When components are deleted from a schematic as a result of an engineering change, the remaining components should not be renumbered.

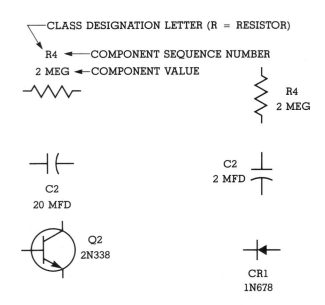

FIGURE 5.5
Reference designation locations.

A more detailed example of a schematic drawing with recommended assignment and location of reference designations is shown in Figure 5.7.

5.5   COMPONENT VALUES

Unit values for reference designations or abbreviations should not be repeated on a schematic diagram. Instead, a general note specifies only the electronic numerical value. Here is a recommended form for the note:

Unless otherwise specified:

Resistance values are in ohms.
Capacitance values are in microfarads.
Inductance values are in microhenries.

For those unit values on a schematic diagram that are exceptions, the value is identified as shown in Figure 5.8.

FIGURE 5.6
Flow of reference designations.

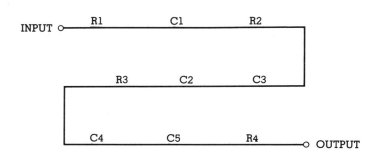

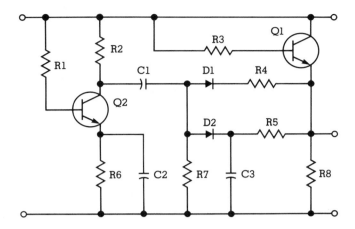

FIGURE 5.7
Assignment and location of reference designations.

FIGURE 5.8
Components with special values.

5.6 METHOD FOR DRAWING SCHEMATIC DIAGRAM

There is no specific or required method for planning and drawing a schematic diagram. There are, however, logical, recommended steps you can take to produce a quality drawing in a minimal amount of time.

- Using light-weight lines, space the major components of the schematic, leaving enough room for each component part as well as for its associated reference designation and part value. The electronic symbols template is used for this portion of the drawing.
- Lightly draw the vertical and horizontal conductor paths that connect the components.

- Examine the layout at this point for completeness, accuracy of information, satisfactory placement of components, and spacing requirements.
- If satisfied, "heavy-up" or darken the symbols and conductor paths using the appropriate line weights.
- Add lettering, reference designation information, component values, and any necessary notes to the drawing.

5.7 SUMMARY

The key to producing schematic diagrams is the drafter's familiarity with electronic graphic symbols. These symbols graphically represent the function of the components of a specific circuit and are identified by an appropriate reference designator. Various line weights and types of lines are used in drawing schematic diagrams to emphasize specific flow path characteristics. Although there is no specific method for planning and drawing the schematic diagram, a series of logical steps helps produce a quality diagram.

Practice exercises for this section are on pages 217 through 231.

5.8 REVIEW EXERCISES

1. What is the purpose of a schematic diagram?

2. What is the direction of the functional flow of information on a schematic diagram?

3. What graphics symbols template is recommended for use by those industries working under government contract?

4. What line weight should be used for general purpose conductor paths on a schematic diagram? For bus bar and cable routing?

5. What is the purpose of reference designations on a schematic diagram?

6. Discuss the recommended placement of reference designations in relationship to graphic symbol location on a schematic diagram.

7. What are the class designation letters for each of the following components?

CAPACITOR _____

FUSE _____

GENERATOR _____

ANTENNA _____

MICROPHONE _____

METER _____

TRANSISTOR _____

SWITCH _____

RECTIFIER _____

8. Given the following reference designation codes, describe the component.

TB1 ——————————————— Q3 ———————————————

R1 ——————————————— K2 ———————————————

C9 ——————————————— P17 ———————————————

T4 ——————————————— J5 ———————————————

S2 ——————————————— F1 ———————————————

CB2 ——————————————— M1 ———————————————

9. Identify the following graphic symbols.

   ———————————————

   ———————————————

   ———————————————

   ———————————————

   ———————————————

———————————————

———————————————

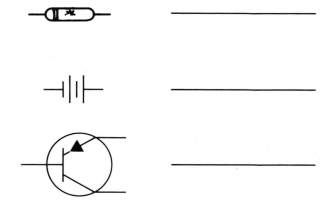

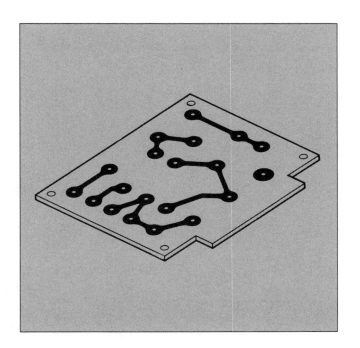

Section 6

THE PRINTED CIRCUIT BOARD

LEARNER OUTCOMES

The student will be able to:

- Draw and calculate the area of electronic components

- Lay out a printed circuit pattern from a schematic

- Produce an artwork drawing

- Define the characteristics of a board detail drawing

- Produce a marking drawing

- Produce a printed wiring assembly drawing

- Demonstrate learning through successful completion of practice and review exercises

FIGURE 6.1
Printed circuit and board.

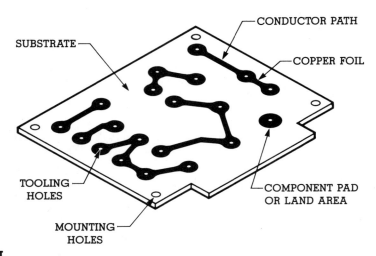

6.1 PURPOSE AND FUNCTION

A printed circuit is a conductive pattern consisting of printed wiring and printed component parts formed on an insulated common base material. When the pattern is located on a completely processed board, it is called a *printed circuit board* (PCB). The purpose of the board is to physically support and electronically interconnect components and parts assembled on it. The function of the assembly is to serve some predetermined purpose. It may operate independently or be part of an electronic system or subsystem.

The printed circuit board is an integral part of a process that leads ultimately to the production of electronic assemblies. The manufacturing process is usually automated and is an approach to mechanized precision production of equipment. This method is the most common process of achieving miniaturization for which there is an ever-increasing demand in the design, development, and manufacture of military and consumer products. Figure 6.1 identifies a printed circuit on an insulated base material.

There are a number of drawings and layouts required for developing and documenting a printed wiring or printed circuit board assembly. The major documents include:

- The circuit board layout drawing
- The artwork drawing, also called the master pattern drawing
- The board detail drawing, also referred to as the master drawing
- The marking drawing, also referred to as the silk-screen drawing
- The assembly drawing, including a list of materials

6.2 THE CIRCUIT BOARD LAYOUT

The development of a printed circuit board assembly begins with an undimensioned board layout developed

from a schematic, a component parts list, a circuit board outline, and other information pertaining to performance and space requirements and limitations of the final package. The layout is a combination of electronic, electrical, or electromechanical components configured to scale on a conductor pattern that meets

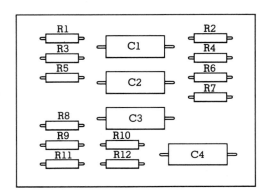

(a) SINGLE DIRECTION ARRANGEMENT

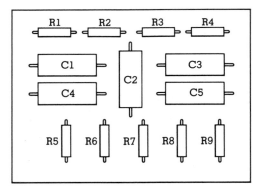

(b) TWO DIRECTION ARRANGEMENT

FIGURE 6.2
Acceptable component orientation: (a) Single-direction arrangement; (b) Two-direction arrangement.

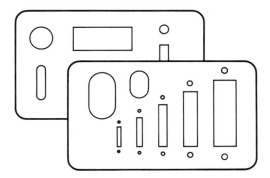

FIGURE 6.3
Component outline template.

some previously established schematic requirement. The components to be mounted are oriented in conformance with the conductor pattern and should ideally be arranged in one direction, but certainly not more than two (see Figure 6.2). The purpose of the layout is to establish proof that the circuit board design is correct and accurate relative to (a) the space requirements between components, (b) the feasibility of the conductor pattern, and (c) the component mounting holes.

The first step for the drafter in preparing the layout is to verify that the board will accommodate all the component parts. There are several methods for determining size verification.

- Method 1   Calculate the approximate area of each component using the maximum body dimensions (length and width, diameter, and height), making certain to allow sufficient space for component leads and to prevent hot spots resulting from excessive power density. Add the areas of all the components and compare this total to the total usable board area.
- Method 2   Using component templates similar to those shown in Figure 6.3, draw outlines of all the components in their approximate locations on the layout. These templates are produced in various scales including 4:1, 2:1, and 1:1.

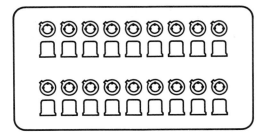

FIGURE 6.4
Component stick-ons (appliqués).

- Method 3   Using commercially available preprinted component outlines, arrange the "stick-ons" in their tentative locations on the board. These component outlines can be rearranged quickly and easily. Figure 6.4 shows examples.

Each of the three methods outlined provides the drafter with the total area that can be used to approximate the minimum circuit board area for components and a rough idea as to final board size. A more precise final board size and component orientation can only be determined with a layout that also considers the conductive pattern.

Dimensionally stable drawing material with a 0.050- or a 0.10-inch modular grid pattern should be used for the initial layout and for drawing, to a 4:1 or 2:1 scale. Scales of 20:1 or 10:1 are used when size or accuracy requirements are critical. The components are laid out on the board according to the schematic drawing and the engineer's design requirements. Whenever possible, land areas, mounting holes, registration and tooling holes, conductive areas, and the board outline should be located on grid intersections or dimensioned from holes located on grid lines. Although this layout is not normally a required drawing, it is basic and essential to all other drawings in the printed circuit board process.

Figure 6.5 shows a schematic provided by an engineer and a flat circuit board layout of all the required components on a 0.10-inch grid.
Figure 6.6 lists the materials to be included in the board layout shown in Figure 6.5, so the drafter can identify the physical size of the components and then verify that the board is large enough.

6.3 CIRCUIT BOARD LAYOUT REVIEW

After completing the circuit board layout and before proceeding with the artwork drawing, the drafter should review the design using the following checklist. An affirmative answer is necessary to the following questions.

1. Are component lead mounting holes on grid intersections whenever possible?
2. Are common mounting lengths correct, including the bend allowance for leads?
3. Are heat sensitive components sufficiently spaced from heat dissipating components?·
4. Are all board dimensions a multiple of the basic grid dimensions whenever possible?
5. Are components mounted so as to be accessible for servicing?
6. Is the board adequately supported?
7. Are adjustable components and test points readily accessible?

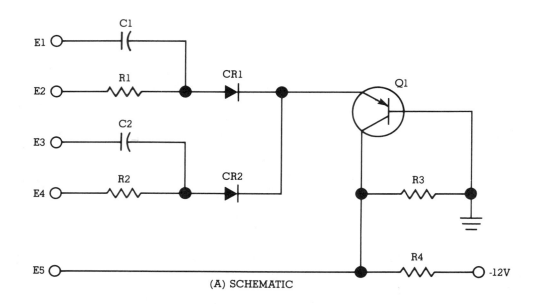

(A) SCHEMATIC

COMPONENT SIDE

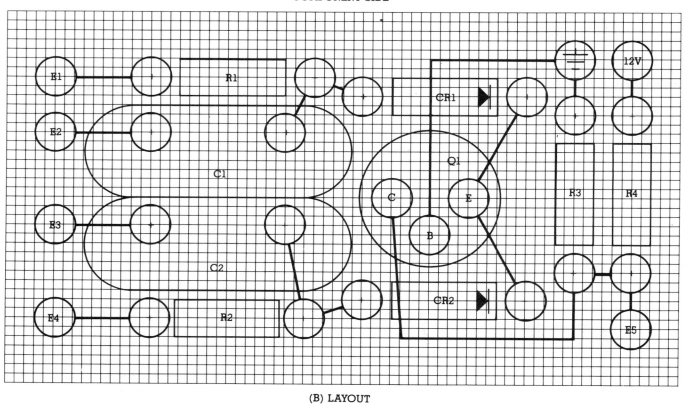

(B) LAYOUT

FIGURE 6.5
Typical flat board layout and related schematic.

LIST OF MATERIALS				

TITLE			DR BY	DATE	DWG NO.	SHEET
CIRCUIT ASSEMBLY					3-001100	1 OF 1

QTY	ITEM	REFERENCE DESIGNATION	PART NUMBER	DESCRIPTION
1	1	R1	3-001120	RESISTOR - 220 OHMS
1	2	R2	3-002120	RESISTOR - 10K OHMS
1	3	R3	3-003120	RESISTOR - 1M OHMS
1	4	R4	3-004120	RESISTOR - 330 OHMS
1	5	C1	4-001120	CAPACITOR - 47 MFD
1	6	C2	4-002120	CAPACITOR - 22 MFD
1	7	D1	5-001120	DIODE - 1N100
1	8	D2	5-002120	DIODE - 1N200
1	9	Q1	6-001120	TRANSISTOR - 2N1304

FIGURE 6.6
List of materials.

8. Are heavy components adequately supported by clamps, clips, or other means?
9. Are conductor paths as short as possible?
10. Does the wiring layout agree with the schematic?
11. Has space been considered for identification of components and test points?
12. Is the cross-sectional area of conductors sufficient to accommodate current (electrical) requirements?

6.4   THE ARTWORK DRAWING

The artwork or the master pattern drawing is a precise scale pattern used to produce the printed wiring or printed circuit within the accuracy specified on the drawing. Care should be taken to produce artwork of the highest quality because, to a large degree, the quality of the artwork directly influences the quality of the printed circuit board.

Normally drawn at a 2:1 or 4:1 scale, the artwork is prepared from the board layout as the second step in the printed circuit layout and design process. Dimensionally stable drafting material such as 0.007-inch thick mylar on a preprinted format of a nonreproducible grid pattern is the most common material for this purpose. The pattern, consisting of conductor paths, land

areas, pads, and terminal areas is represented by black photo-opaque, pressure sensitive appliqués that can be cut to form a pattern, or can take the form of a die-cut shape to fit a specific purpose. Examples of some of the more common shapes and their uses are displayed in Figure 6.7.

Conductor lengths between terminal areas should be held to a minimum. The route should be as short and straight as possible, as shown in Figure 6.8.

You should avoid conductor corners that have less than a 90-degree included angle, because foil delamination may occur. Exterior corners should have a radius as illustrated in Figure 6.9.

Terminal areas, also referred to as land areas, should have smooth profiles that consist of simple arcs as shown in Figure 6.10. A terminal area is provided for each point of electrical attachment of a component part lead or other electrical connection to the printed circuit board. The minimum diameter for terminal areas is determined by the required size of mounting holes or type of connection.

The artwork pattern must exhibit good workmanship: sharpness and uniformity of lines, uniform spacing, and smoothness of curves; with no visible voids, pinholes, discontinuities, or other imperfections.

You must usually place three registration marks on the artwork so as to form a 90-degree angle. The purpose of registration marks is to align the several

FIGURE 6.7
Common appliqué shapes.

DONUT PAD

CONDUCTOR - STRAIGHT

OVAL PAD

TEAR DROP PAD

TRIANGULAR PAD

HEX PAD

SQUARE PAD

CONNECTOR PAD

CONDUCTOR - RADIUS

MULTIPURPOSE

DUAL IN-LINE PAD

FLAT PACK

drawings so that when they are overlaid, they will be accurate from one sheet to the next. Figure 6.11 shows the most common registration mark, also called a *target* because of its shape.

In addition, a photographer's reduction dimension is placed either above or to the side of the artwork. This dimension is an important aid to the photographer for reducing the artwork to its proper size, as

seen in Figure 6.12. Board boundary markers, also called *trim marks*, are approximately 0.125 wide, with the inside edge representing the circuit board's length and width dimensions. The marks are placed at the corners and are made with conductor path tape.

6.5 ARTWORK DRAWING REVIEW

When the artwork is complete the drafter should review it by asking the following questions. All these points should be satisfied before going further.

1. Does conductor spacing meet the minimum requirements?

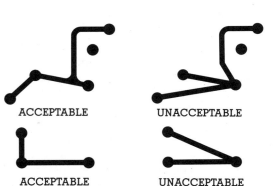

ACCEPTABLE   UNACCEPTABLE

ACCEPTABLE   UNACCEPTABLE

FIGURE 6.8
Conductor lengths.

ACCEPTABLE

UNACCEPTABLE

FIGURE 6.9
Conductor corners.

SOLDER WILL NOT FLOW
EVENLY TO FILLET

SOLDER WILL FLOW AWAY
FROM FILLET

SOLDER FILLET WILL BE
NON-SYMMETRICAL

OUTSIDE SOLDER FILLETS
WILL BE NON-SYMMETRICAL

CENTER SOLDER FILLET
WILL BE POOR

SOLDER WILL FLOW
TOWARD LARGER LAND

ACCEPTABLE   UNACCEPTABLE

FIGURE 6.10
Terminal area profiles.

2. Do the conductive pattern, numerals, letters, and characters have dense, uniform, black color with smooth edges?
3. Have sharp corners in the pattern been avoided?
4. Are fillets and radii properly utilized?
5. Do all the appropriate processing marks appear on the artwork? Are they positioned accurately?
6. Does the pattern reflect good workmanship?
7. Have special design considerations been met?

6.6 THE BOARD DETAIL DRAWING

This drawing is also called the *master drawing* and is reproduced from the artwork drawing. It contains sufficient information so that the engineering department can check the circuit and verify fabrication. For this reason this drawing is also referred to as an engineering drawing or fabrication drawing. The completed drawing identifies the size and shape of the board, size and

FIGURE 6.11
Registration mark or target.

location of all holes, location of keyways, if required, and the shape or arrangement of both conductive and nonconductive patterns or elements. All features not controlled by the artwork are located by dimensions or specified by notes that include material and plating requirements, and any other information pertinent to fabricating and assembling the package. An example of a board detail drawing is shown in Figure 6.13, page 69.

6.7 THE MARKING DRAWING

The marking drawing is also referred to as the *silk-screen drawing* because of the process normally used to place component identifiers on the circuit board. The purpose of the marking drawing is to identify each component that is to become part of the printed circuit board assembly. The drawing is produced to scale and accurately indicates the location of all components. In some cases the contour or component outline is shown on the drawing; in other cases only component identification is shown, without the outline. A combination of component outlines, identifiers, and additional nomenclature such as terminal numbers, special information, and the name of the subassembly are included, as shown in Figure 6.14, page 70. Note that this drawing is always viewed from the component side of the circuit board. The drawing is usually produced by using the artwork drawing as a template, component side up, so that the required component identifiers can be positioned in the desired proximity.

6.8 THE ASSEMBLY DRAWING

The printed circuit board assembly drawing depicts the completed board and the location and mounting of all electronic, electromechanical, and mechanical components including clamps, clips, or other component retaining devices. This drawing is made to the same scale as the board detail drawing and is also viewed from the component side of the board. An accompanying parts list and any necessary assembly instructions are also considered part of the drawing. At times, supplemental views or sections may be necessary to show additional dimensions, part location, part orientation, or assembly sequence. Figure 6.15 on page 71 is a typical printed wiring assembly drawing. Figure 6.16 on page 72 shows an assembly drawing that requires supplemental information.

6.9 SUMMARY

The printed circuit board and its associated components represent a series of drawings that relate directly to the fabrication and manufacture of electronic and electromechanical assemblies. The major drawings include (a) the circuit board layout drawing, (b) the artwork drawing, (c) the board detail drawing, (d) the marking drawing, and (e) the assembly drawing. Several drafting aids and materials help produce high qual-

ity drawings quickly and accurately. We have outlined several acceptable and unacceptable graphics techniques and shown examples of each type of drawing.

Practice exercises for this section are on pages 233 through 255.

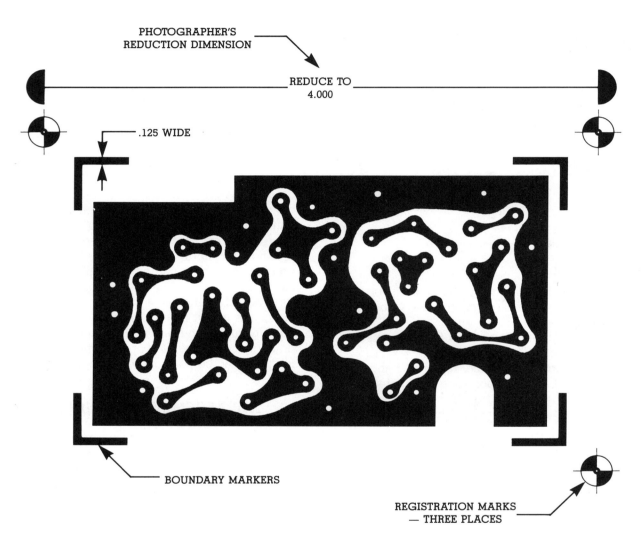

FIGURE 6.12
Artwork and markings.

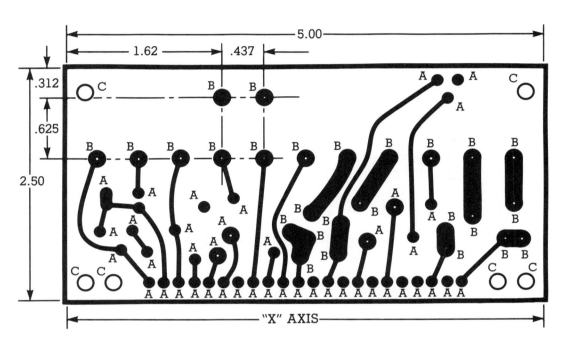

HOLE CHART		
HOLE	DESCRIPTION	QTY
A	.070	42
B	.086	26
C	.188	6

NOTES:
1. MAT'L. - .054 TO .071 THICK G-10 WITH
   2 OZ. COPPER ON ONE SIDE ONLY.
2. FINISH - 60/40 TIN-LEAD IMMERSION.
3. PACKING - SEAL IN PLASTIC.
4. ARTWORK - SEE SHEETS 3 & 4 FOR TOOLING
   INFORMATION.
5. ALL HOLES NOT LOCATED BY DIMENSIONS
   ARE LOCATED ON 0.010 GRID INTERSECTIONS
6. MAXIMUM TWIST OR WARP ALONG "X" AXIS
   IS .090 FOR THIS AREA.
7. SEE E-49912 FOR COMPONENT MOUNTING
   DIMENSIONS.

DRAWN		
CHECKED		DETAIL, BOARD
ENGINEER		
DATE	APPROVED	
CONTRACT NO.	SIZE	DRAWING NO.
	A	6602 AKT
ENG. RELEASE		
	SCALE FULL	SHEET 1 OF 1

FIGURE 6.13
Board detail drawing.

COMPONENT SIDE

CONTROL
INTERFACE   K1   C1 C2

                                          R2   C3 ▭
R14   R4   C4 ▭
      TIME   R5
      DELAY   P1   R6   Q1
R11   R7

1   3

2   4   TB1   R3   R1

P2   1 2 3 4 5 6 7 8   ADJ.   ADJ.
                 DRIVE FOLLOWER CARD

DRAWN		DRAWING, MARKING	
CHECKED			
ENGINEER			
DATE	APPROVED		
CONTRACT NO.		SIZE	DRAWING NO.
ENG. RELEASE		A	5602 AKT
		SCALE	SHEET 1 OF 1

FIGURE 6.14
Marking drawing.

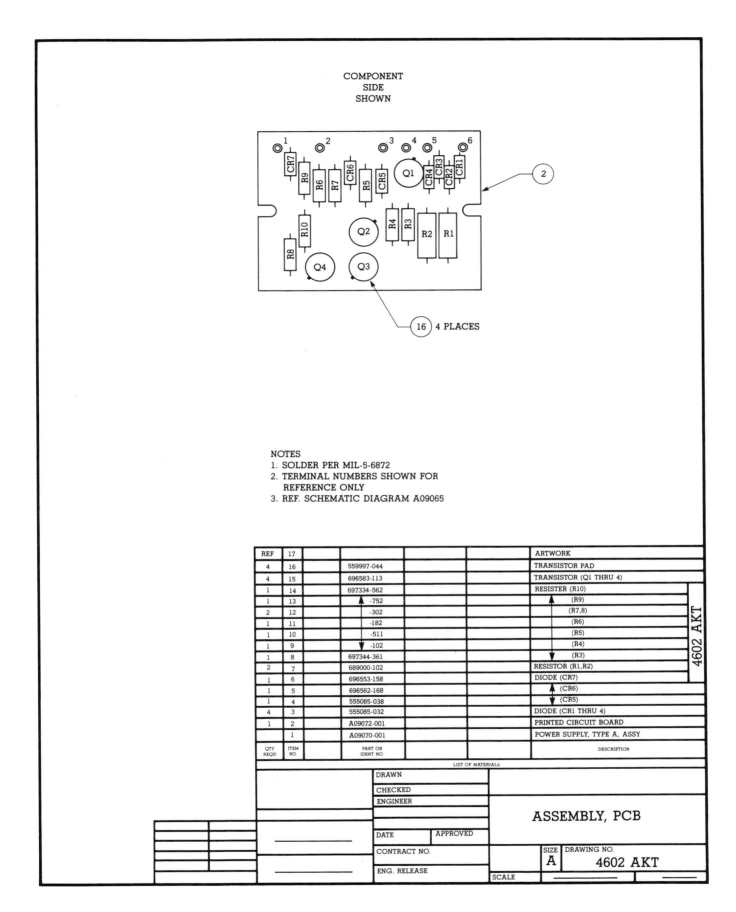

COMPONENT
SIDE
SHOWN

NOTES
1. SOLDER PER MIL-5-6872
2. TERMINAL NUMBERS SHOWN FOR
   REFERENCE ONLY
3. REF. SCHEMATIC DIAGRAM A09065

QTY REQD	ITEM NO.		PART OR IDENT NO.			DESCRIPTION	
REF	17					ARTWORK	
4	16		559997-044			TRANSISTOR PAD	
4	15		696583-113			TRANSISTOR (Q1 THRU 4)	
1	14		697334-562			RESISTER (R10)	
1	13		-752			(R9)	
2	12		-302			(R7,8)	
1	11		-182			(R6)	
1	10		-511			(R5)	
1	9		-102			(R4)	
1	8		697344-361			(R3)	
2	7		689000-102			RESISTER (R1,R2)	
1	6		696553-158			DIODE (CR7)	
1	5		696562-168			(CR6)	
1	4		555085-038			(CR5)	
4	3		555085-032			DIODE (CR1 THRU 4)	
1	2		A09072-001			PRINTED CIRCUIT BOARD	
1	1		A09070-001			POWER SUPPLY, TYPE A, ASSY	

LIST OF MATERIALS

DRAWN			ASSEMBLY, PCB	
CHECKED				
ENGINEER				
DATE	APPROVED			
CONTRACT NO.			SIZE A	DRAWING NO. 4602 AKT
ENG. RELEASE			SCALE	

4602 AKT

FIGURE 6.15
Assembly drawing.

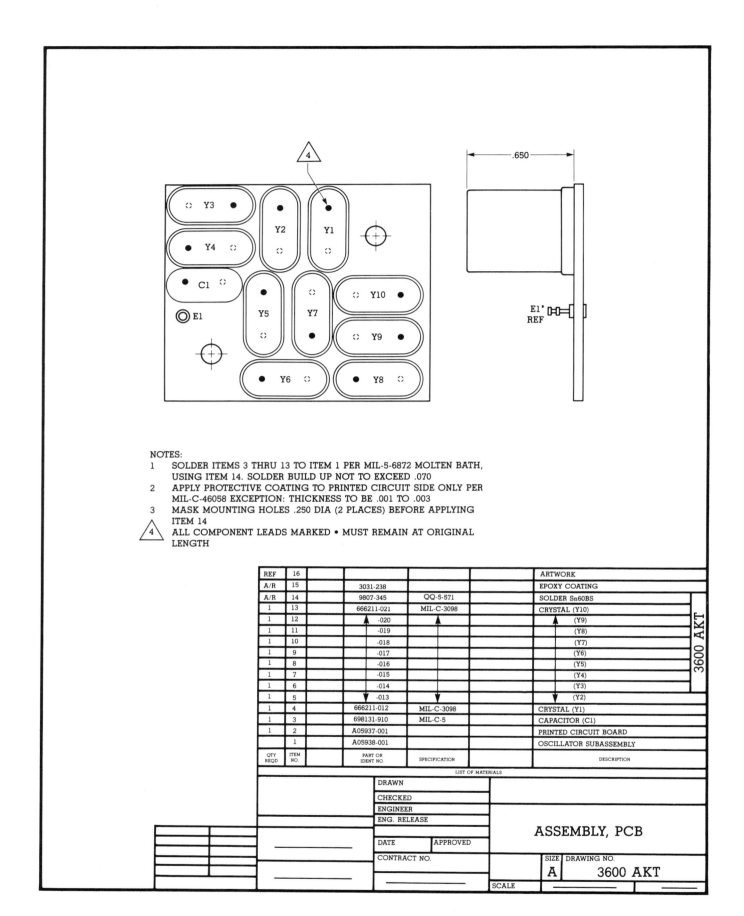

FIGURE 6.16

Assembly drawing with supplemental data.

6.10 REVIEW EXERCISES

1. List, in proper sequence, the major drawings required to produce a printed circuit board assembly.

2. Explain two methods of assuring that the circuit board will accommodate all the required component parts.

3. What stable drawing material is normally used for the initial printed circuit board layout?

4. What is the purpose of a circuit board layout?

5. What material is normally used to depict land areas and conductor paths on the artwork drawing?

6. List five questions a drafter needs to answer after completing the layout and before starting the artwork.

7. Why should the artwork drawing be of high quality?

8. Draw five common shapes of land areas, pads, and terminal areas used in laying out printed circuit boards.

9. What is the purpose of registration marks? Boundary marks? Photographer's reduction dimension?

10. What is another name for the board detail drawing?

11. To what scale is the printed circuit assembly drawn?

12. List three items of information included in a completed board detail drawing.

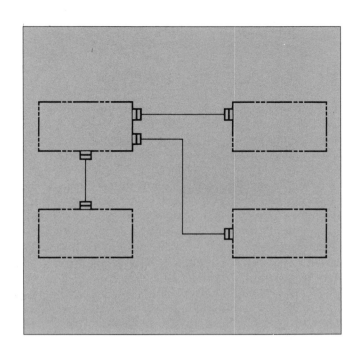

Section 7

THE INTER-CONNECTION DIAGRAM

LEARNER OUTCOMES

The student will be able to:

- Identify reference designations on an interconnection diagram

- Describe the purpose and function of an interconnection diagram

- List three types of interconnection diagrams

- Complete an interconnection diagram from incomplete data

- Demonstrate learning through successful completion of practice and review exercises

7.1 PURPOSE AND FUNCTION

The interconnection diagram can be prepared at any level of drawing where two or more electrical assemblies are combined. This diagram is a form of wiring diagram that shows, schematically, all functional interconnections among assemblies and usually identifies only external connections among these units. It also supplies reference information to other drawings and diagrams including appropriate mechanical assembly data. "Inter" means "between," and "connection" means "to join," so for our purpose, interconnection means "joining or connecting between" units or assemblies. Internal connections within units are normally omitted, but for clarity, partial internal connection information is sometimes provided on this type of diagram.

The function of the interconnection diagram is to portray simply the "to" and "from" connections between units that take the form of shapes such as squares, rectangles, and circles connected by lines identified by reference designations. The interconnection diagram is a useful document for tracing an area or system circuitry from one major item to another and is frequently used for troubleshooting. It can also be used as a planning guide for establishing interconnecting cabling requirements. The reference document used by government contractors and major commercial industries when producing interconnection diagrams is the American National Standards Institute (ANSI) Y14.15a specification.

7.2 TYPES OF INTERCONNECTION DIAGRAMS

Three types of interconnection diagrams can each be produced by different methods. The *wiring diagram* shows each wire; the *cabling* type shows cables only; and the *tabular* type normally takes the form of a wire or running list. (We will discuss the tabular type of diagram in Section 8.) In this section we will discuss the wiring, continuous line point-to-point type, and the cabling, diagrammatic type.

7.2.1. CONTINUOUS LINE POINT-TO-POINT TYPE

Continuous line point-to-point diagrams show an individual connecting line between each pair of terminals to be connected. Each connecting line is identified with one or more of the following:

- Wire color information
- Wire size (area)
- Wire type (voltage rating, construction, shielding, specification)

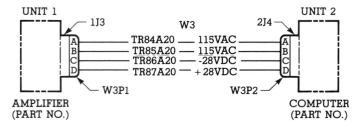

FIGURE 7.1
Wiring type, continuous line point-to-point interconnection diagram.

Information describing the destination of the wires and the circuit function of the connection may also be included. These identifications are placed in breaks in the connecting line, as illustrated in Figure 7.1.

7.2.2 CABLING, DIAGRAMMATIC TYPE

The cabling diagram may be prepared as a continuous line type. Each cable is represented by a single line running between two or more items connected by a cable. Connectors, terminal boards, and the like are represented by single line graphic symbols. An example of this type of diagram appears in Figure 7.2.

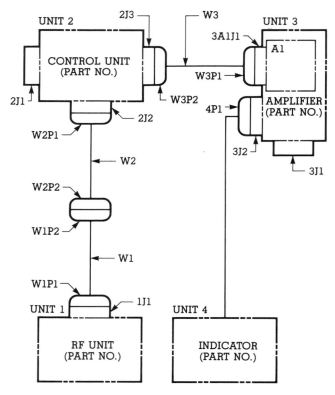

FIGURE 7.2
Diagrammatic form, cabling interconnection diagram.

7.3 LAYOUT OF THE DIAGRAM

To lay out the diagram, one normally places all units or assemblies on the outer periphery of the drawing and shows them as phantom-lined figures. When space permits, connectors that mate with external equipment should appear at the left or right side of the diagram. Assemblies or units pertinent to the diagram will appear at the top or bottom of the drawing, as shown in Figure 7.3. Views should be shown as though all connections are in one plane. Figure 7.3 also shows that interconnecting paths between assemblies and circuitry integral to the assembly level of the diagram are in the center portion of the drawing. To simplify the drawing and to minimize the number of crossover paths, assemblies should be placed next to one another. For clarity, circuitry that is an integral part of an assembly may be shown. In Figure 7.3, for example, switch (S1) and meter (DS1) are located immediately adjacent to connecting components or assemblies. On the diagram, the electrical function of each interconnection is identified. The interconnection paths (lines) are also defined and the electrical function inserted in the broken lines.

Figure 7.4 illustrates a cable type interconnection diagram for a personal computer system that allows for transmission of data through a telephone line. All cables, jacks, plugs, and assemblies are identified. For clarification, identification of the reference designators are shown in Figure 7.4 as follows:

J = jack or connector (female connection)
P = plug (male connection)
W = cable

In addition, each assembly is identified by a UNIT NUMBER and FUNCTIONAL TITLE, where:

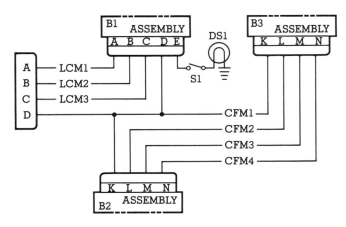

FIGURE 7.3
Wiring type, continuous line interconnection diagram layout.

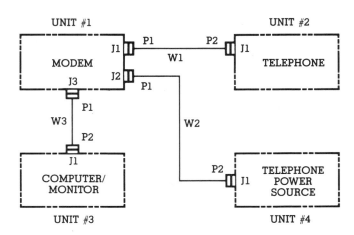

FIGURE 7.4
Cabling type, interconnection diagram.

UNIT #1 = MODEM
UNIT #2 = TELEPHONE
UNIT #3 = COMPUTER/MONITOR
UNIT #4 = TELEPHONE POWER SOURCE

Therefore, in Figure 7.4, 3J1 identifies UNIT #3 (COMPUTER/MONITOR), and its associated female jack (J1), while W2 (cable) is the connecting cable between 1J2 (MODEM) and 4J1 (TELEPHONE POWER SOURCE).

7.3.1 CONNECTING LINES

On interconnection diagrams, the same convention for connecting lines should be used as on schematic diagrams. Conductor paths should be drawn only horizontally or vertically. Parallel lines should be spaced at intervals of not less than 1/16-inch when the diagram is reduced to its final size. Crossover paths should be kept to a minimum. Connection dots should be drawn as a solid circle of 3/32-inch diameter, as we saw in Figure 5.4.

7.4 METHOD FOR DRAWING THE INTERCONNECTION DIAGRAM

There are no specific rules for drawing the interconnection diagram, nor are there required standard symbols for this type of diagram, but drawing templates are available and should be used when possible. The following steps will help assure a quality diagram:

1. Collect data from the engineer, designer, or supervisor of the project. The data should include one or more of the following: (1) a schematic diagram; (2) a layout drawing; (3) a manufacturer's drawing; and (4) if available, a wire running list.

2. Determine the correct size and type of drawing paper to accommodate the drawing and its associated information.
3. Using light-weight lines, lay out the major assemblies or units of the diagram, keeping in mind that connectors that mate with external equipment should be placed at the left and/or right side of the drawing and the major units in the upper and lower portion of the diagram. This allows space in the central part of the drawing for interconnecting paths (lines).
4. Lightly draw the vertical and horizontal interconnection lines between units, assemblies, and internal and external connectors.
5. Review the drawing at this point for completeness of information, spacing of items, and accuracy.
6. If satisfied, "heavy-up" all lines, leaving sufficient space for cable, connector, and/or wire identification information. Use medium line weight and maintain consistency.

7. Add all remaining information such as reference designations, lettering, identification of electrical functions, and cable information.

7.5 SUMMARY

The interconnection diagram is a form of wiring diagram that indicates the functional interconnections between assemblies. We discussed three types of diagrams: (1) the wiring type; (2) the cabling type; and (3) the tabular type. The interconnection diagram is used to trace area or system circuitry, for troubleshooting, and as a planning aid for establishing cabling requirements. Specific details for diagram layout and a step-by-step method for drawing the diagram were introduced.

Practice exercises for this section are on pages 257 through 263.

7.6 REVIEW EXERCISES

1. Name three types of interconnection diagrams.

2. Name two industrial uses for the interconnection diagram.

3. What geometric shapes are normally used to identify units or assemblies?

4. What do the following terms indicate?

P =

W =

J =

5. Describe the difference between the wiring type and the cabling type of interconnection diagram.

6. When space permits, where should connectors that mate with external equipment appear on an interconnection diagram?

7. What data must a drafter have before beginning an interconnection diagram?

8. What line weight should be used for initially laying out the diagram?

9. When should the drafter "heavy-up" the lines of the diagram?

10. Describe the line convention used for producing an interconnection diagram.

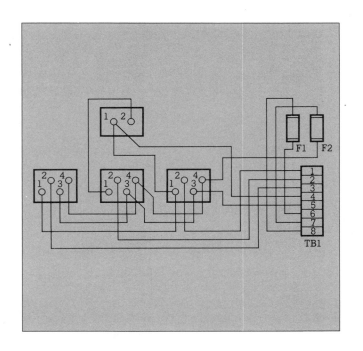

Section 8

THE CONNECTION DIAGRAM

LEARNER OUTCOMES

The student will be able to:

- Identify several types of connection diagrams

- Describe the characteristics of the point-to-point, highway, and tabular type wiring diagrams

- Identify conductor path designation information for a wiring diagram

- Produce several different types of connection diagrams

- Define the purpose and use of a wiring harness diagram

- Identify the difference between the conventional cable and flat cable assemblies

- Demonstrate learning through successful completion of practice and review exercises

8.1 INTRODUCTION

Several types of diagrams and drawings are required for producing electronic or electromechanical assemblies. Connection diagrams provide production line personnel with the "how to" for wiring connections from unit to unit or assembly to assembly. Three types of connection drawings are used for this purpose: (1) the wiring diagram, (2) the cable assembly drawing, and (3) the wiring harness diagram.

8.2 THE WIRING DIAGRAM

8.2.1 PURPOSE AND FUNCTION

As we saw in Section 7, the interconnection diagram shows external wire connections and cabling between units and/or assemblies. The wiring diagram, which is a form of connection diagram, differs from the interconnection diagram because it may identify both the internal and external connections of an item such as an electromechanical assembly, a piece of equipment, or a unit. The wiring diagram's purpose is to show the general physical size and location of all component parts of an assembly. It contains enough detailed information to enable one to trace all the electrical connections involved.

The wiring diagram has several uses in a production facility. It is used for troubleshooting, maintenance of equipment, and quality control, as well as for modifying equipment. This type of diagram is also used to train workers in point-to-point wiring techniques. It is prepared at the drawing level for which a mechanical assembly and circuit diagram are prepared.

8.2.2 TYPES OF WIRING DIAGRAMS

Industry uses several types of wiring diagrams, including the point-to-point type, the highway type, base line, and the straight line type. In addition, the wiring diagram is often produced in tabular form as a running list or as a computer printout, also called a wire list. We will discuss the point-to-point, highway, and tabular type diagrams in this section.

8.2.2.1 The Point-to-Point Diagram

The point-to-point type wiring diagram is a form of continuous line drawing where lines represent the actual connections between various components. This type of diagram shows those items to be wired in a simple outline form on the wiring side of the unit and correctly oriented in their approximate location, as illustrated in Figure 8.1. The connecting lines or wires are shown as individual lines drawn either horizontally or vertically (with a few exceptions for clarity), and the physical re-

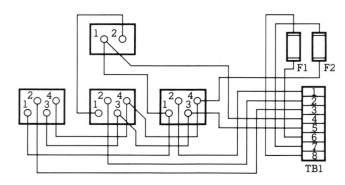

FIGURE 8.1
Point-to-Point type wiring diagram.

lationship between components is not necessarily accurate; that is, components are not in the precise positions shown on the assembly drawing of the unit for which the wiring diagram is to be produced.

8.2.2.2 The Highway Type Diagram

The highway type wiring diagram, also referred to as the *trunkline wiring diagram,* is portrayed in Figure 8.2. We can see that several individual lines are grouped together to make a single path called a *highway* or *trunkline.* Short feeder lines connect the terminals of the components with the highway line and also show the direction of the wire path. More than one highway may be shown on the diagram depending on the physical location of the components and the route of the conductor paths (wires).

Critical wiring, however, should not be included as part of the highway line, but drawn as individual lines. It is the responsibility of the engineer or designer to determine which wires are critical. In addition, entry and exit lines on the diagram are drawn at a 45-degree angle to indicate wire direction. Wire identification is placed next to the entry and exit points to clearly define the path and termination point of each wire. More often than not, a break appears in the conductor path to accommodate wire identification.

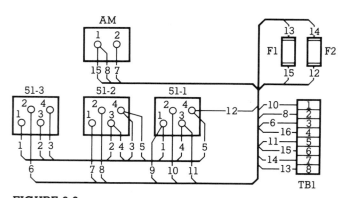

FIGURE 8.2
Highway type wiring diagram.

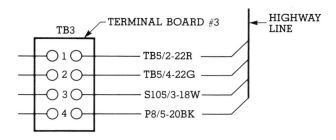

FIGURE 8.3
Conductor path designation information.

Major industries conform to an American National Standards Institute (ANSI) method of presenting conductor path designation information on diagrams. This information includes (a) component designation, (b) component part or terminal, (c) wire size or type, if required, and (d) wire color. Figure 8.3 illustrates the designation of information from terminal board 3 (TB3) to the highway, which is shown as a heavy line. In the illustration, TB5/2-22R indicates that TB5 is the destination of the wire to TERMINAL BOARD #5; 2 is the terminal connection at the destination; −22 is the wire size at the destination, TERMINAL BOARD #5; and R is the color of the wire, RED. Appendix D lists the Color Coding and Color Abbreviations as recommended by American National Standards Institute (ANSI) Standard—Y14.15.

Symbols for components on the diagram are located in approximately the same position as the components are located in the equipment. Within reason, placement may differ from the actual location to avoid possible interference or crowding of connecting lines or designations. Symbols take the shape of circles, squares, or rectangles, as illustrated in Figures 8.1 and

8.2, to approximate the shape of the component or of a terminal. The size of the symbol need not be exact, but should be of such proportion and detail to identify the component. The size of the symbol also depends on the final reduction size of the diagram. Symbols are drawn in heavier line weights than the conductor paths that join the symbols.

8.2.2.3 Tabular Type Diagram
On the tabular type diagram, wiring information is arranged in a table called a wire data list or running list, rather than drawn as a line diagram. This drawing consists of printed data and the necessary instructions to establish wiring connections within an assembly or system. The wire data list identifies the destination and termination requirements for each wire of a multiconductor harness or cable. Figure 8.4 shows the information that appears on a typical tabular drawing. This list may be shown on the same sheet as the diagram, on a separate sheet of the same drawing, or as a separate drawing. Figure 8.5 illustrates how the wire data list is shown as part of the drawing.

8.2.3 DIAGRAM NOTES

A wiring diagram is complete only after all the necessary general or specific data relative to wire size and color or processes are added. Examples of typical notes may include:

a. UNLESS OTHERWISE NOTED, ALL WIRES TO BE #16 AWG AND UNDERWRITER LABORATORY APPROVED.
b. SHIELDED WIRE SHOWN TO BE #20AWG.
c. SOLDERING SHALL BE IN ACCORDANCE WITH SPECIFICATION S135.

FIGURE 8.4
Tabular type wiring data.

WIRE DATA LIST					
WIRE NO.	GAUGE	COLOR	STOCK NUMBER	FROM	TO
10A	18	BK	82134-001	B1	B8
10B	20	BR	82135-003	B2	B12
11A	16	R	82136-005	B2	B3
11B	22	O	82137-007	B4	B17
11C	18	Y	82138-009	B6	B11
12A	24	G	82139-011	B7	B9
13A	20	BL	82140-012	B5	B13
					B14

8.2.4 METHOD FOR DRAWING THE WIRING DIAGRAM

1. Gather all the necessary information and printed data for beginning the drawing. This information includes the schematic diagram and the electromechanical assembly drawing from which the connection drawing can be produced. The schematic diagram is necessary for correctly tracing component paths from component to component. The assembly drawing helps to verify the schematic and to locate on the wiring diagram all the wired components that make up the assembly.

2. Depending upon the complexity of the data, the contractural requirements for the project, or the engineering leader's instructions, determine the necessary type of wiring diagram (i.e., point-to-point, highway, tabular, etc.).

3. Select the appropriate drawing material for the diagram, (i.e., vellum, mylar, grid, or plain) as well as the correct size drawing material (i.e., 11-inch × 17-inch, 22-inch × 34-inch, etc.). Remember to arrange the wiring diagram on the drawing format so that it will read from left to right, keeping the input on the left and output on the right. Each additional layer, if required, should read the same way.

4. Using drafting instruments and light-weight lines, draw the component symbols in the form of squares, rectangles, or circles, observing the general outline of the component. Put them in the approximate locations identified on the assembly drawing.

5. Special Note: Before lightly drawing in the connection lines, note that the lines connect one terminal to another or to an electrical or mechanical tie point, not to junctions or connections as on sche-

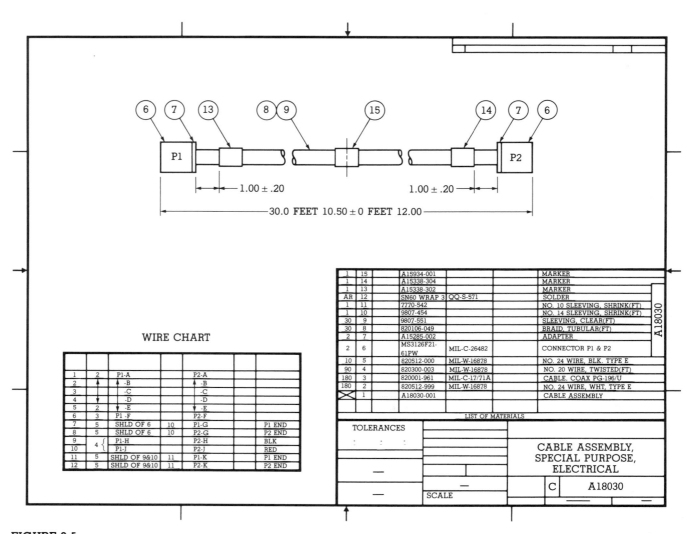

FIGURE 8.5
Typical cable assembly drawing.

matic diagrams. Draw only vertical and horizontal lines, except for angled short lines on a highway drawing. Try to keep connection lines equally spaced, if possible.

6. Examine the drawing at this point for proportion, completeness of information, and satisfactory placement of symbols. Be sure to allow sufficient space to accommodate reference designation information, notes, and a wire data list, if they are to become part of the diagram.

7. If satisfied, "heavy-up" or darken the symbols, connection lines, and highway lines according to these guidelines: (1) connection lines are medium weight, (2) symbols are heavy, and (3) highway lines are extra-heavy (to show that the highway line is a bundle of wires, not just a single wire).

8. Add conductor path designation information and any general or special notes relative to wire size, color, and process information. Include a wire data list, if required, with all the necessary "to" and "from" information.

8.3 THE CABLE ASSEMBLY DRAWING

8.3.1 PURPOSE AND FUNCTION

The cable assembly drawing depicts a definite length of insulated conductor or conductors with one or more ends processed or terminated by connectors or fittings that allow connection to other items. Figure 8.5 illustrates a typical cable assembly drawing for a company under military contract; several military specifications (MS) are identified. The assembly provides reference to the following information: overall dimensions and tolerances, preparation of the ends of the cable (P1, P2), chart of wires, identifying wires, termination points, sleeving types, markers, and list of materials.

Figure 8.6 shows a cable assembly whose configuration differs from that shown in Figure 8.5. The cable in Figure 8.6 is called a *flat* or *ribbon cable,* a usually thin and very flexible cable that takes a minimal amount of space. It can be used where a lot of physical movement is required in the equipment, such as be-

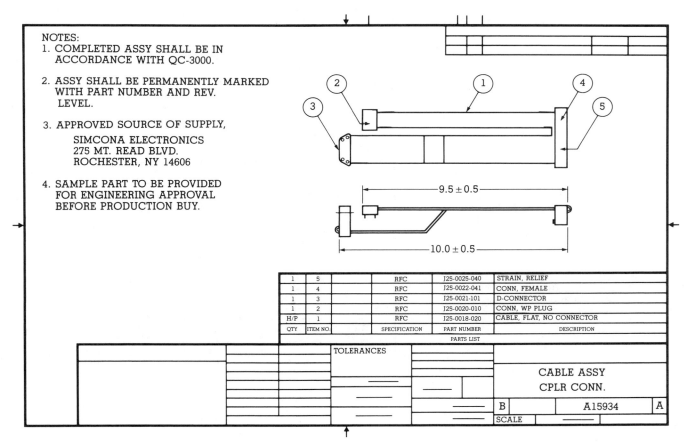

FIGURE 8.6
Flat cable assembly.

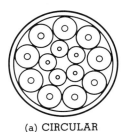

(a) CIRCULAR

(b) FLAT (RIBBON)

FIGURE 8.7
Cable assembly cross section: (a) Circular; (b) Flat (ribbon).

hind chassis drawers or in hinged areas. Figure 8.7 illustrates the major difference between the two types of cable assemblies. The conventional cable is a round bundle of conductors, whereas the conductors in the flat ribbon cable lie side by side. In addition, to accommodate the round bundle of wires, a round multipin connector is used for the conventional cable assembly, while the ribbon cable requires a flat in-line connector.

8.3.2 DIMENSIONS AND TOLERANCES FOR CABLE ASSEMBLY DRAWINGS

The overall length of a cable assembly is dimensioned from the ends of straight connectors and from the outside edge of an angled connector, as shown in Figure 8.8. Length dimensions are normally shown in inches to these generally approved manufacturing tolerances:

Length of cable	Tolerance
0 to 10 inches	± .50 inches or ± 10% (whichever is greater)
10 to 34 inches	± 1.00 inch
34 to 72 inches	± 2.00 inches
72 to 120 inches	± 4.00 inches
120 to 240 inches	± 6.00 inches
240 to 600 inches	± 10.00 inches
over 600 inches	± 12.00 inches

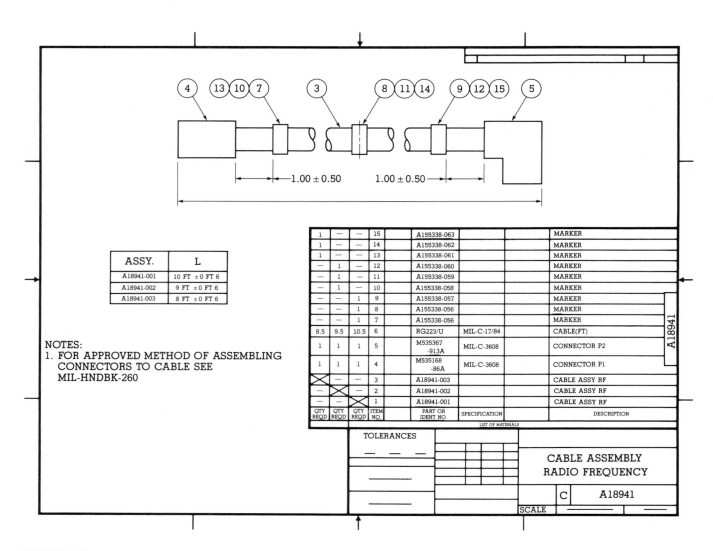

FIGURE 8.8
Cable assembly dimensioning practices.

8.4 THE WIRING HARNESS DIAGRAM

The wiring harness diagram is also known as a *cable form diagram*. It is an undimensioned drawing that portrays a cable form and termination points of individually insulated conductors arranged in paths and bound together by lacing twine, plastic clips, or other similar bindings. The shape and contour of the harness is usually determined by examining the preproduction model of the electromechanical unit, subassembly, or assembly for which the wiring harness is being developed. The wiring harness diagram is drawn in a single plane and includes all the information necessary to determine the cable form and its termination points. The diagram is normally drawn full-scale because it is used as a template or guide for producing the harness assembly. Technicians or assembly workers frequently use the diagram to produce the first production model

of a harness. An example of a wiring harness diagram is shown in Figure 8.9.

One way the diagram is used by those producing the actual harness assembly is depicted in Figure 8.10. The process is as follows:

1. A copy of the harness drawing is glued to a piece of wood, normally ½-inch or ¾-inch thick plywood. The length and width dimensions of the wood depend on the size of the full-scale diagram.
2. Nails are driven into the board at appropriate locations along the outline of the drawing, equally spaced (as shown), to make certain that when the individual conductor paths are followed, the final harness assembly conforms to the diagram.
3. Nails or dowel pins are also placed at branch or end points where sufficient conductor path slack is allowed to make the necessary connections.

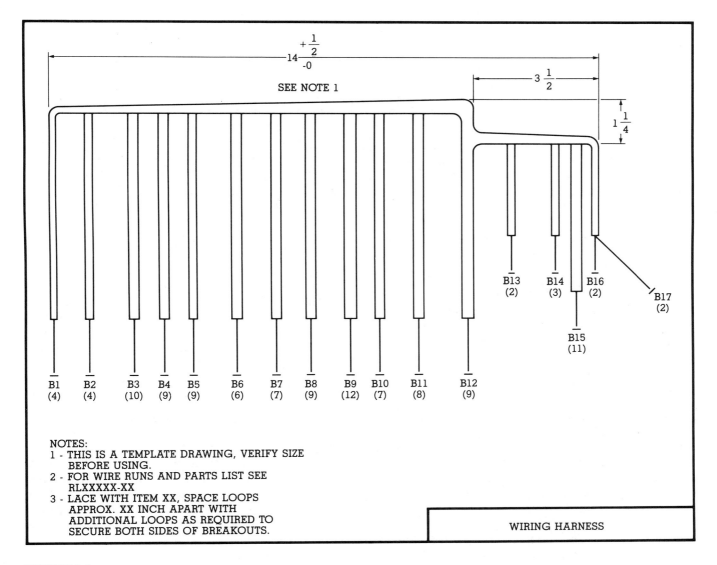

FIGURE 8.9
Wiring harness diagram.

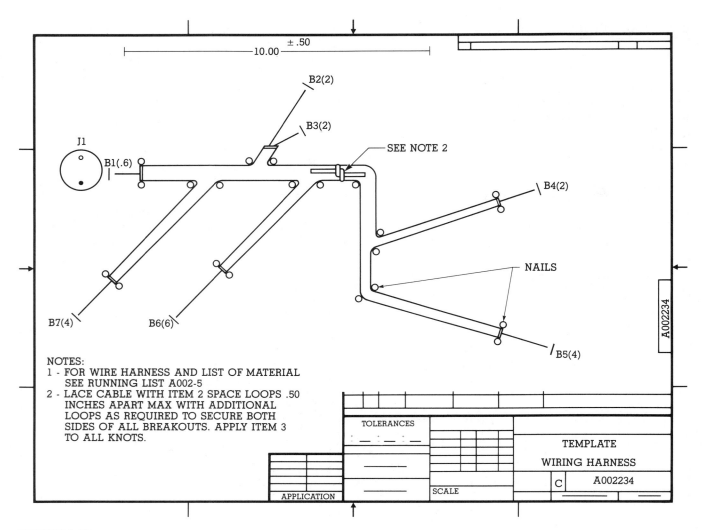

FIGURE 8.10
Use for a wiring harness diagram.

8.4.1 DIMENSIONING

Although the harness diagram is called an undimensioned drawing, it may require an overall dimension to provide a check of the print size before it is used as a manufacturing template. Other dimensions may be needed to control the size and shape of the harness illustrated in Figure 8.9.

8.4.2 BREAKOUT NUMBERS

A wiring harness diagram may not be complete in itself. Wire destinations are identified and listed on an associated wire running list according to the wire breakouts from the laced portion of the harness diagram. Wires in a harness may also be tagged for identification with marker sleeves (hollow plastic tubing called *spaghetti*). A *breakout* is that portion of the wiring harness that splits off or leaves the main bundle of conductors. Each breakout is designated by a letter followed by an identifying number. In Figure 8.9, B1 through B17 are examples of wire breakout designations.

Wire lengths are indicated by cut lines in Figure 8.11, where lacing and a breakout number are also shown. Lacing is either shown on the diagram or in a note, as in Note 3 on Figure 8.9.

8.4.3 RUNNING (WIRE) LIST

The running list is a book-form drawing that includes tabular data and instructions for establishing wiring connections within an assembly or between any combination of assemblies or units of a system. A running list is a form of wiring diagram similar to that shown in Figure 8.4 as a wire data list. The running list document may be used as a computer input document or as an officially released engineering drawing.

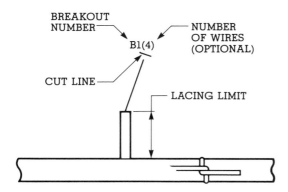

FIGURE 8.11
Wire cut line, lacing, and breakout number.

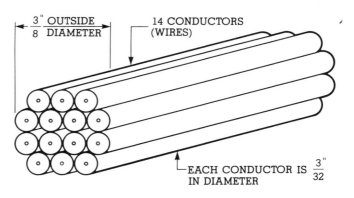

FIGURE 8.12
Cable diameter approximation.

8.4.4 METHOD FOR DRAWING THE WIRING HARNESS DIAGRAM

- Try to obtain the actual physical assembly for which the harness diagram needs to be developed. With the actual assembly at hand, it is easier to measure and scale the necessary lengths and bends required for the harness.
- If the physical assembly is not available, a full-scale copy of the mechanical assembly drawing will provide enough information to trace the general path the harness should take.
- Attach a print of the assembly drawing to the drafting board. Over this, lay the appropriate size translucent drawing paper. Develop the basic outline of the harness using light-weight lines. The diameter of each conductor may vary, but as a guide, use $3/32$-inches as the average diameter per conductor. Figure 8.12 illustrates 14 conductors in a bundle measuring approximately $3/8$-inches in diameter.
- After the major bundle of wires is laid out, determine location and length of breakout points.
- If satisfied that all diagram lines for the harness have been placed on the drawing, "heavy-up" the harness outline.
- Identify each wire with its appropriate associated information.
- Produce a complete running list that includes the item number of the wire, the wire gage of each con-

ductor, the part number, the wire color, and the "to" and "from" information for each wire. (The running list may or may not be part of the wiring harness diagram.)
- As a check on the quality of the harness layout, it is good practice to fabricate the complete harness and install it in a prototype before releasing final drawings.

8.5 SUMMARY

Various industrial standards apply to the production of drawings for connection diagrams. We reviewed the wiring diagram, the cable assembly drawing, and the wiring harness diagram in this section. All require special drawings that may include color coding of conductors, identification of conductor types and sizes, routing information, point-to-point connection data, and component designation. Symbols for components take the form of circles, squares, and rectangles on a drawing, depending upon the component's physical outline. We have outlined methods for producing both the wiring diagram and wiring harness diagram.

Practice exercises for this section are on pages 265 through 277.

8.6   REVIEW EXERCISES

1.   Name three types of connection diagrams.

2.   What purpose does the wiring diagram serve in a production facility?

3.   List five types of wiring diagrams.

4.   How do the point-to-point and the highway wiring diagrams differ?

5.   How are the "feeder" lines shown on the highway wiring diagram?

6.   From Figure 8.3, define the meaning of:
TB5/4-22G

_____   _____
_____   _____

S105/3-18W

_____   _____
_____   _____

P8/5-20BK

_____   _____
_____   _____

7.   What is the preferred abbreviation for these conductor colors?
Blue _____   Brown _____
Orange _____   White _____
Gray _____   Yellow _____

8.   What information is typically shown on a wire data list (tabular form)?

9. What information is necessary before you can start a wiring diagram? Why?

10. What line weight should be used on a wiring diagram for these items?
    conductor path
    components
    highway lines

11. Name five items of information that are identified on a cable assembly drawing.

12. What is the generally approved tolerance for a cable assembly that is 84 inches long?

13. What are the characteristics of a wiring harness diagram?

14. Explain how a wiring harness diagram is used in industry.

15. What is a "breakout" on a wiring harness diagram?

16. What is a running list?

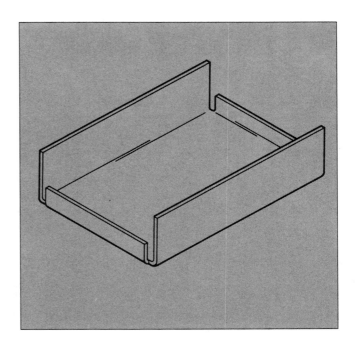

Section 9

ELECTRO-MECHANICAL PACKAGING

LEARNER OUTCOMES

The student will be able to:

- Define electromechanical packaging

- List the project responsibilities for the designer and drafter

- Calculate clearance hole sizes and grip length for fasteners used in electronic devices

- Calculate the total developed length of a bracket with 90-degree bends

- Identify the steps in developing an electromechanical package

- Create and produce several drawings for an electromechanical project

- Demonstrate learning through successful completion of practice and review exercises

9.1 INTRODUCTION

Electronics industries form a significant part of manufacturing plants throughout the United States. Many large firms that formerly produced primarily mechanical products have diversified their product lines to include electronic and electromechanical devices. There has also been a proliferation of computer and related peripheral products in recent years. As a result, electromechanical packaging is in demand for items ranging from as large as complete military systems for aerospace, shipboard, or ground use, to microminiaturized units for integrated circuits.

The three types of manufactured products are classified as military, consumer, and commercial items. Each product may require adherence to different design, reliability, human factors, maintainability, and environmental specifications. The variability in these criteria must receive appropriate attention in their design.

In the organization of tasks in an engineering department, design of electronic devices usually includes the packaging of electronics parts. *Packaging* refers to designing, developing, and/or drafting a case, enclosure, or package to house electronic components in such a way as to fulfill the intended customer's requirements for size and shape, and perhaps also considering environmental and reliability factors. Because the package may be constructed of metal and non-metal parts, the complete assembly will probably consist of electronic as well as mechanical items, hence the term *electromechanical packaging*. Therefore, designers and drafters, whose major tasks occur at the drafting board, perform electromechanical packaging tasks.

9.2 DESIGNER/DRAFTER RESPONSIBILITIES

Tasks for the designer/drafter include both electronic and mechanical design. The designer is charged with the conceptual, developmental, and design work along with some drafting that includes establishing the physical layout of the equipment to be designed. The designer's responsibility may vary from industry to industry, depending on organizational structure, size of the engineering department, and the type of manufactured products. The designer may work directly for an electronics engineer, a project engineer, or a drafting/design supervisor, and is usually a member of a project team. A thorough knowledge of materials, processes, and shop procedures for the manufacture of electronic equipment is a strong asset to the designer.

The drafter may also be a member of a project team that might include several drafters working with the same designer or a number of designers depending upon the complexity and size of the project. The drafter's responsibility does not normally include as much design work as that of a designer, but rather more drafting of detail parts, subassembly and assembly drawings, and perhaps also drawing block diagrams, logic, schematic, connection and interconnection diagrams, cable and harness diagrams, component drawings, and printed circuit board design and layout.

9.3 TYPES OF EQUIPMENT ENCLOSURES

Equipment enclosures and mountings for electronic components can take many forms. The terms *rack*, *panel*, *cabinet*, and *chassis* are commonly used in the industry. These items can be defined as follows:

■ Rack—A standard rack is a rectangular metal structure that consists of long vertical members. It houses electronics equipment mounted in drawers with slides, or permanently mounted by other means behind a panel. A standard electronics rack is illustrated in Figure 9.1.

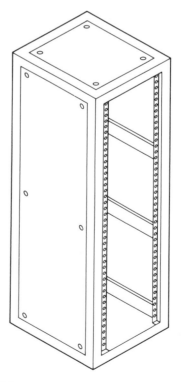

FIGURE 9.1
Standard electronics rack.

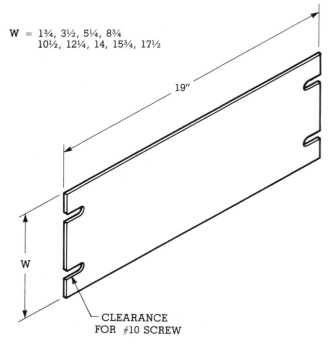

W = 1¾, 3½, 5¼, 8¾
10½, 12¼, 14, 15¾, 17½

19″

W

CLEARANCE
FOR #10 SCREW

FIGURE 9.2
Typical rack-mounted panel.

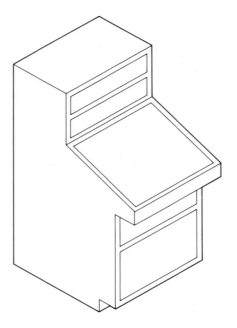

FIGURE 9.3
Cabinet or console.

- Panel—A standard rack-mounted panel is a rectangular metal plate, usually 19 inches long and in heights of 1¾-inch increments. A rack-mounted panel is designed to fit a standard rack and is usually fastened to the rack by screws. Figure 9.2 shows a typical rack-mounted panel.
- Cabinet—A cabinet is a metal structure of either a standard or special shape, depending on the application. Throughout the electronics industry, the maximum established height for cabinets is 86 inches. The horizontal mountings are designed to accept a standard rack-mounted panel. A cabinet with a desk-top surface for an operator work area may be referred to as a *console*. Figure 9.3 shows a cabinet.
- Chassis—A chassis is a thin-gage metal structure or base that supports electronic, electromechanical, or mechanical components. It may be of any size or shape and can be mounted to a standard rack panel or used separately. It may or may not have a cover or be enclosed.

One of the most common operations performed on a chassis is forming or bending. This operation takes place after all holes, slots, and cutouts are made in the blank, or when the chassis is in the form of a flat piece of thin-gage metal. Bending can be accomplished on a hand- or power-operated machine called a brake. The quantity of chassis to be produced deter-

mines whether a hand brake or a power-operated one is used.

The radius of a bend varies depending on the material and its thickness. If too small a radius is specified on a drawing, the material may fracture at the bend. You can use Appendix G, BENDS—RADII IN SHEET METALS for reference when designing a chassis. In addition, when drawing a thin-gage mechanical chassis, cover, or bracket, the fully developed length and width need to be determined so that material can be obtained and the department producing the part will have an idea as to the initial size of stock to be used.

Appendix G, BENDS—90 DEGREE DEVELOPED LENGTH may be used as reference when calculating developed lengths with 90-degree bends.

Figure 9.4 shows two types of chassis design. Figure 9.4(a) is a U-shaped chassis and (b) is a box type. The box type can be used upright or inverted depending on the need. Determining the type of chassis to use on a particular project depends on many factors: the number of components to be mounted, weight of the heaviest components, circuit and operator protection, location of controls, and environmental factors.

Equipment enclosures can be made of several different materials, such as high-impact sheet or molded plastic, steel, aluminum, and aluminum alloy stock. Thin-gage metal, shapes, angles, castings, and extrusions (specially formed shapes) are commonly used to meet the needs of a packaging project.

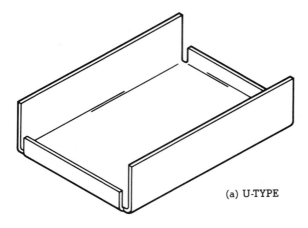

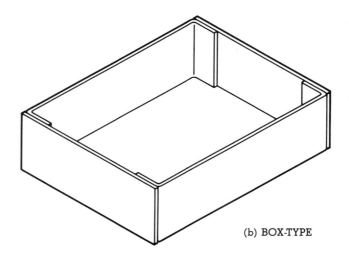

(a) U-TYPE

(b) BOX-TYPE

FIGURE 9.4
Common chassis types: (a) U-type; (b) Box-type.

9.4 THE CHASSIS LAYOUT

The designer usually draws the chassis layout, often from a rough sketch of a component layout provided by the design engineer. Space limitations usually determine the maximum size for a chassis. Ideally, additional information provided by the responsible engineer includes a schematic diagram to support the rough component layout and a preliminary parts list and/or existing drawings of all components expected to be used in the design. A chassis detail drawing for an electronics project might look like the one in Figure 9.5.

Placement of components on a chassis layout is determined by many factors including freedom from electrical and mechanical interference, ease of acces-

sibility, and provision for testing and maintenance of the assembly. Templates or a paper cutout of components are often used in place of the actual components on the layout to determine the best possible design.

9.5 DRAFTING PRACTICES

The technical graphics practices outlined in Section 1 should be followed for producing quality chassis layout and detail work. You will need to place special emphasis on line convention, orthographic projection, and dimensioning practices. Because the chassis is very often made of thin-gage aluminum or steel, it is helpful for the drafter to be familiar with Decimal Equivalents

FIGURE 9.5
Chassis detail drawing.

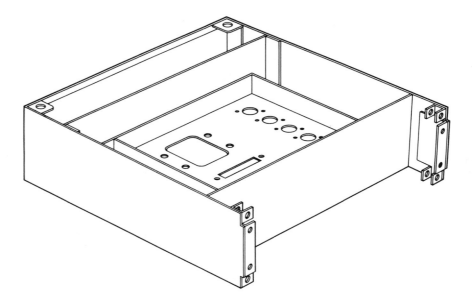

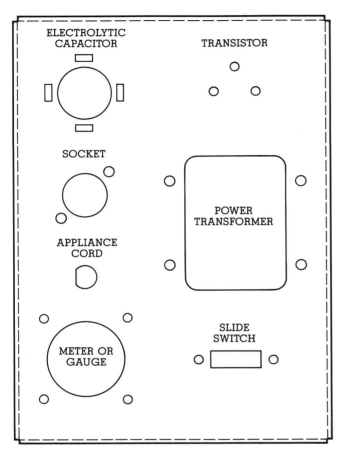

FIGURE 9.6
Common cutout shapes.

of Wire and Sheet gages, listed in Appendix E, page 147.

Several sizes and shapes of cutouts and holes are used in developing a chassis. Some of the more common shapes are included in Figure 9.6.

9.6 FASTENING METHODS

In the fabrication process, several methods are used for fastening electronic and mechanical components and parts together. The methods include welding (both spot and heliarc), brazing, soldering, riveting, staking, bonding with epoxies, and fastening with screws. The most commonly used screws are the pan head, the flat head, the hex socket head, and the hex head machine screw. They are identified by their military standard (MS) base number, along with washers and hex nuts, in Table 9.1.

Commercially available specialty fasteners are also often used for electronics assembly. They include devices that require only a one-quarter turn to fasten securely and do not fall out during assembly or in use. Several types of fasteners can be tightened by hand instead of requiring a tool.

The following practice is used to determine the size of clearance holes and tolerances for hole locating dimensions when two or more fasteners are used to join mating parts. We will describe four common conditions with an example of each. In the accompanying formulas the notations are:

TABLE 9.1
Standard hardware sizes and dimensions.

SCREW		PAN HEAD		FLAT HEAD		SOCKET HEAD		LOCK WASHER MEDIUM		FLAT WASHER		HEX HEAD MACHINE						HEX NUT					
SIZE NO.	DIA	A	H	A	H	A	H	OD	T	OD	T	A	H	P	A	H	P	A	T	P	A	T	P
2	.086	.167	.062	.172	.051	.140	.086	.175	.026	.260	.025	.125	.050	.144				.187	.066	.217			
4	.112	.219	.080	.225	.067	.183	.112	.212	.031	.260	.028	.187	.060	.216				.250	.098	.289			
6	.138	.270	.097	.279	.083	.226	.138	.253	.037	.385	.065	.250	.080	.288				.312	.114	.361			
8	.164	.322	.115	.332	.100	.270	.164	.296	.046	.385	.065	.250	.110	.288				.343	.130	.397			
10	.190	.373	.133	.385	.116	.313	.190	.337	.053	.448	.065	.312	.120	.360				.375	.130	.433			
1/4	.250	.492	.175	.507	.153	.375	.250	.493	.072	.635	.080				.437	.163	.505				.437	.226	.505
5/16	.312	.615	.218	.635	.191	.469	.313	.591	.088	.697	.080				.500	.211	.577				.500	.273	.577
3/8	.375	.740	.261	.762	.230	.563	.375	.688	.104	.822	.080				.562	.243	.650				.562	.337	.650

PREFERRED BASE NO.: MS51957 MS51958 | MS51959 MS51960 | MS16995 MS16996 | MS35338 | MS15795 | 639123 | MS35307 MS35308 | MS35649 MS35650 | MS35690

M = maximum screw diameter
T = sum of the bilateral tolerance between holes
D = minimum diameter of clearance hole

These examples are for a No. 8 pan head screw (.164 diameter) located by a dimension with a tolerance of + .010 inches, − .010 inches (T = .020).

■ *Condition No. 1:* Pattern of two holes with clearance holes in both parts.

$$M + \frac{T}{2} = D$$

$$.164 + \frac{.020}{2} = .174 \text{ Diameter}$$

Closest hole is .177 (No. 16 drill)

■ *Condition No. 2:* Pattern of two holes with clearance hole in one part and tapped or countersunk holes in mating part.

$$M + T = D$$
$$.164 + .020 = .184 \text{ Diameter}$$

Closest hole size is .185 (No. 13 drill)

■ *Condition No. 3:* Pattern of more than two holes with clearance holes in both parts.

$$M + 1.4\,T = D$$
$$.164 + 1.4\,(.020) = .192 \text{ Diameter}$$

Closest hole size is .1935 (No. 10 drill)

■ *Condition No. 4:* Pattern of more than two holes with clearance hole in one part and tapped or countersunk holes in mating part.

$$M + 2.8\,T = D$$
$$.164 + 2.8\,(.020) = .220 \text{ Diameter}$$

Closest hole size is .221 (No. 2 drill)

Table 9.2 is arranged to show the nearest drill size for conditions 1–4 for pan head screws.

TABLE 9.2
Clearance holes for pan head screws.

SIZE OF SCREW	CONDI-TION	± .005 DRILL SIZE	± .010 DRILL SIZE	± .015 DRILL SIZE	± .020 DRILL SIZE	± .025 DRILL SIZE	± .030 DRILL SIZE
NO. 2 (.086 DIA.)	1	.0935(42)	.096(41)	.1015(38)	.1065(36)	.111(34)	.116(32)
	2	.096(41)	.1065(36)	.116(32)	.1285(30)*,#	.136(29)*,#	.147(26)*,#
	3	.1015(38)	.116(32)	.1285(30)*,#	.144(27)*,#	.157(22)*,#	.173(17)*,#
	4	.116(32)	.144(27)*,#	.173(17)*,#	.199(8)*,#	.228(1)*,#	.257(F)*,#
NO. 4 (.112 DIA.)	1	.120(31)	.125(1/8)	.1285(30)	.136(29)	.1405(28)	.144(27)
	2	.125(1/8)	.136(29)	.144(27)	.152(24)*	.166(19)*	.173(17)*,#
	3	.1285(30)	.1405(28)	.154(23)*	.1695(18)*,#	.182(14)*,#	.196(9)*,#
	4	.1405(28)	.1695(18)*,#	.196(9)*,#	.228(1)*,#	.257(F)*,#	.2811(K)*,#
NO. 6 (.138 DIA.)	1	.144(27)	.1495(25)	.154(23)	.159(21)	.166(19)	.1695(18)
	2	.1459(25)	.159(21)	.1695(18)	.180(15)	.189(12)*	.199(8)*
	3	.152(24)	.166(19)	.180(15)	.196(9)*	.209(4)*,#	.228(1)*,#
	4	.166(19)	.196(9)*	.228(1)*,#	.250(1/4)*,#	.2811(K)*,#	.3125(5/16)*,#
NO. 8 (.164 DIA.)	1	.1695(18)	.177(16)	.180(15)	.185(13)	.189(12)	.196(9)
	2	.177(16)	.185(13)	.196(9)	.204(6)	.2187(7/32)	.228(1)*,#
	3	.180(15)	.1935(10)	.209(4)	.221(2)*	.2344(15/64)*	.250(1/4)*,#
	4	.1935(10)	.221(2)*	.250(1/4)*,#	.277(J)*,#	.3125(5/16)*,#	.332(Q)*,#
NO. 10 (.190 DIA.)	1	.196(9)	.201(7)	.2055(5)	.213(3)	.2187(7/32)	.221(2)
	2	.201(7)	.213(3)	.221(2)	.234(A)	.242(C)	.250(1/4)
	3	.204(6)	.2187(7/32)	.234(A)	.246(D)	.261(G)	.277(J)*
	4	.2188(7/32)	.246(D)	.277(J)*	.302(N)*,#	.332(Q)*,#	.3594(23/64)*,#
1/4" (.250 DIA.)	1	.257(F)	.261(G)	.266(H)	.272(I)	.277(J)	.2811(K)
	2	.261(G)	.272(I)	.2811(K)	.290(L)	.302(N)	.3125(5/16)
	3	.2656(17/64)	.2811(K)	.295(M)	.3125(5/16)	.323(P)	.339(R)
	4	.2811(K)	.3125(5/16)	.339(R)	.368(U)*	.3906(25/64)*	.4219(27/64)*,#

NOTE - DRILL SIZE IS CLOSEST SIZE WITHOUT TOLERANCE CONSIDERATION
* MEDIUM LOCKWASHER WILL DROP INTO CLEARANCE HOLE UNDER WORST CONDITION
# HEAD OF SCREW WILL NOT COVER HOLE UNDER WORST CONDITION

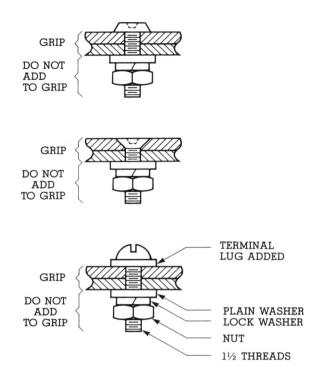

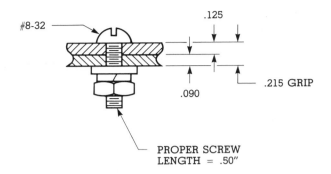

FIGURE 9.8
Grip length calculation.

FIGURE 9.7
Examples of grip length.

9.6.1 SCREW LENGTH SELECTION

It is a drafter's responsibility to determine length of hardware for such items as rivets, mounting posts, standoffs, and screws. Given the combination of material thickness, washers, and nuts, there are some rules for determining the correct screw length. *Grip* is the sum of any combination of material thickness *plus any*

hardware under the screw or fastener head, as shown in Figure 9.7. *Do not add the thickness of any hardware, or for 1½ threads below the material*, because it has been calculated into the accompanying Screw Length Chart, Table 9.3. To determine proper screw length, calculate the grip as indicated in Figure 9.8. If the calculated grip result is not shown in Table 9.3 under the appropriate screw size, use the next larger grip value. The proper screw length is found on the same line in the column to the extreme left, "Screw Length."

For example, since .215 is not listed under No. 8-32 screw size in Table 9.3, use the next larger grip, which is .236. It specifies a .500-inch screw length. This becomes the correct screw length to use for this particular application. When a washer is deleted, as illustrated in Figure 9.9, subtract its thickness (.049) from the sum of the grip dimension (.215) to obtain a correct grip of .166 inches. Since a grip of .166 is not listed in Table 9.3 under the 8-32 screw column, use

TABLE 9.3
Screw length chart.

SCREW SIZE	2-56	4-40	6-32	8-32	10-32	1/4	5/16	3/8
SCREW LENGTH								
.125	—	—	—	—	—	—	—	—
.188	.056	—	—	—	—	—	—	—
.250	.118	.068	.011	—	—	—	—	—
.312	.180	.130	.073	.048	.041	—	—	—
.375	.243	.193	.136	.111	.104	.012	—	—
.438	.306	.256	.199	.174	.167	.074	—	—
.500	.368	.318	.261	.236	.229	.074	—	—
.625	.493	.443	.386	.361	.354	.199	.128	.039
.750	.618	.568	.511	.486	.479	.324	.253	.164
.875	.743	.693	.636	.611	.604	.449	.378	.289
1.000	—	.818	.761	.736	.729	.574	.503	.414
1.250	—	1.068	1.011	.986	.979	.824	.753	.664
1.500	—	1.318	1.261	1.236	1.229	1.074	1.003	.914
1.750	—	—	1.511	1.486	1.479	1.324	1.253	1.164
2.000	—	—	1.761	1.736	1.729	1.574	1.503	1.414
2.250	—	—	—	1.986	1.979	1.824	1.753	1.664
2.500	—	—	—	2.236	2.229	2.074	2.003	1.914
2.750	—	—	—	2.486	2.479	2.324	2.253	2.164
3.000	—	—	—	2.736	2.729	2.574	2.503	2.414

FIGURE 9.9
Corrected grip length.

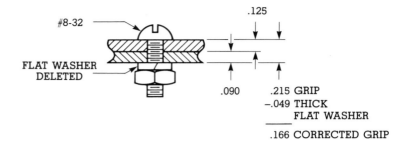

the next larger grip, .174, which specifies a .438 inch screw length. This becomes the appropriate screw length for this particular application.

9.7 METHOD FOR DEVELOPING AN ELECTROMECHANICAL PACKAGE

An electromechanical package or assembly may consist of several parts, including a chassis, cover, panel, printed circuit board, meters, gages, switches, as well as numerous other components. Before the electronics designer/drafter can begin to work on the package, the project leader must provide certain printed documentation, including a schematic diagram, a preliminary parts list, and any rough sketches that may be relative to the mechanical or electronic aspects of the design, such as electromagnetic interference, clearance for adjustable components, and size limitations.

The drawings the designer/drafter is likely to produce for a project may vary, but generally include:

- A chassis layout drawing showing the location of major electronic and all mechanical parts
- Detail drawings of all mechanical parts and electronic component parts
- Subassembly drawings, if required
- A final assembly drawing
- Electronic drawings and diagrams
- A complete bill of materials

This is one recommended process for developing a project:

1. After obtaining all the preliminary data relative to the project, select the appropriate drawing medium, usually vellum, large enough to make the layout full-scale, if possible.
2. Position major electronic components on the layout by drawing or with paper cutouts. Paper cutouts allow you to move components around on the layout until you produce a suitable layout and obtain preliminary approval. Determine the location of the components by considering such factors as accessibility for replacement, short electronic connections, and freedom from interference of mechanical parts or electronic components that need adjustment.
3. After preliminary layout approval, draw center lines for all major components. These will serve as reference points for locating other components. Add the appropriate hole patterns on the chassis for component mountings and fasteners.
4. Make certain that the layout has sufficient information for producing detail drawings.
5. You can now make detail drawings. Complete all detail drawings, including all mechanical parts and electronic components. Calculate sizes, types, and length of hardware and fasteners at this time.
6. Complete the subassembly and assembly drawing, including any notes, special processes, and assembly instructions.
7. Produce all required electronic drawings and diagrams.
8. Develop the bill of material for the assembly. The bill of material may be part of the drawing or may take the form of a separate sheet, possibly computer-generated.

9.8 SUMMARY

Electromechanical packaging may be required for anything from a microminiaturized subassembly to a large, complex military system. The roles of the designer and drafter are significant to the success of the project. They require skill in the area of technical graphics and knowledge of the design of enclosures, methods of fabrication, production techniques, and various types of fastening methods. The various types of drawings they must produce include mechanical, electromechanical, and electronic drawings and diagrams. The designer and drafter are part of a project team that may include several different professional and technical engineering personnel.

Practice exercises for this section are on pages 279 through 297.

9.9 REVIEW EXERCISES

1.   Name three types of electronics products manufactured in this country.

2.   List the responsibilities of the designer in the development of a project.

3.   What are the major tasks of an electronics drafter on a project?

4.   What is the purpose of an electronics rack?

5.   What is a chassis? What is it used for?

6.   List four types of screws used as fasteners on an electromechanical assembly.

7.   Identify five kinds of fastening methods that can be used on an electromechanical package.

8.   For a #10 screw, determine the diameter of the clearance holes required in a pattern of two holes with a bilateral tolerance of .020 inches.

9.   Determine the correct screw length for a #6-32 screw whose grip thickness is .261 inches.

10. Identify the type of information the electronics designer or drafter needs before beginning an electromechanical packaging project.

11. What are four specifications that can have an impact on the design of an electromechanical package?

12. List three steps for the designer/drafter in developing an electromechanical package.

13. Calculate the total developed length for the mounting bracket identified below. Use Appendix G for reference.

14. Calculate the total developed length for an identical bracket to the one shown in problem 13, with the following changes:
a. Inside bend radius = .188
b. Material thickness = .064

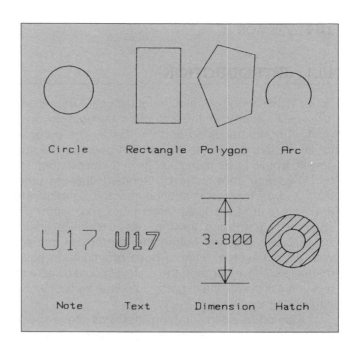

Section 10

INTRODUCTION TO COMPUTER-AIDED DRAFTING SYSTEMS

LEARNER OUTCOMES

The student will be able to:

- List the major hardware elements of a computer-aided drafting (CAD) system and describe their functions.

- Identify the purpose of a CAD system.

- Describe the difference between hardware and software.

- Define several terms of CAD terminology.

- Be aware of the impact of CAD on the production of electronic and electrical diagrams and drawings.

- Produce several electronic and mechanical drawings and diagrams using CAD system hardware and software.

- Demonstrate learning through successful completion of practice and review exercises.

10.1 INTRODUCTION

So far in this worktext, all the educational experiences have been aimed at the student's eventual involvement in computer-aided drafting (CAD). The purpose for all the learning in the course has been for the student to obtain basic manual skills in technical graphics by (1) gaining knowledge of standard drawing practices and drafting room procedures and (2) acquiring the skills to produce quality electronics and electromechanical diagrams and drawings. Basic technical graphics skills are required before the drafter can interact with a CAD system. With a good grounding in basic manual skill development, the drafter should be able to move quickly and intelligently into the field of computer graphics.

Computer-aided drafting systems have been used by government agencies and industry for several years, providing new and imaginative ways to produce drawings. The primary purpose of a CAD system is to increase productivity by enabling the drafter/designer to make neater, cleaner, more accurate drawings in less time. A CAD system is an electronic tool that replaces traditional drafting equipment and tools. The key to any successful CAD system is the operator—in our case, the drafter. Computer-aided drafting systems are being used in all engineering fields, in some to a greater degree than others. One field where it is used extensively is in the production of electronics diagrams and drawings.

Some acronyms are used throughout the computer graphics industry by various companies who design and manufacture computer graphics systems and equipment:

CAD	—	Computer aided design/drafting
	—	Computer-assisted design/drafting
CADD	—	Computer-aided design and drafting
	—	Computer-assisted design and drafting
CAE	—	Computer-aided engineering
CAA	—	Computer-aided artwork
CAG	—	Computer-aided graphics
ADS	—	Automatic drafting system
EGS	—	Engineering graphics system

10.2 BASIC CAD CONCEPT

Let us consider a simplified version of a basic computer-aided drafting (CAD) concept: first, the computer receives information (input from a keyboard or digitizer); second, it processes that information and displays the result (output on the monitor or plotter). For a two-dimensional drafting system, the computer uses data (the input) provided by the operator (drafter) to produce (by processing the data) a drawing (the output). Figure 10.1 illustrates a typical computer-aided drafting concept.

CAD systems are generally interactive. *Interactive* implies an interaction or communication among the various components of the system as well as with the operator. The types of graphics that can be produced accurately and rapidly for electronic-related industries include general mechanical and electromechanical drawings, schematic and logic diagrams, printed circuit board layouts and design (including single and multilayered configurations), connection and interconnection diagrams, block diagrams, running lists, and lists of materials.

Although a time-consuming learning process must take place before acquiring proficiency in the use of CAD hardware and software, some claim that productivity by engineering personnel can be increased by a factor of 4 to 1 over conventional or traditional practices for design and drafting, depending on the type of CAD system and the number of peripheral options. The learning curve for those in training to become CAD operators varies according to many factors, but a period of three to six months of training on a CAD system usually allows the operator to become proficient.

FIGURE 10.1
Basic CAD concept.

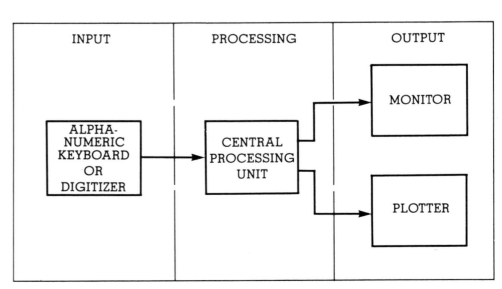

FIGURE 10.2
Two-dimensional interactive
CAD system.

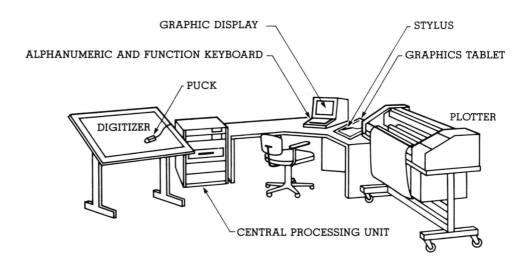

Relatively high costs have prevented educational institutions from procuring large mainframe or turnkey CAD systems, but several firms have recently entered the market with mid-range to low cost systems that use mini- or microcomputers, desk-top computers with work stations and "user friendly," intelligent terminals. We will describe several of these systems and define their structures, capabilities, similarities, and differences.

10.3 SYSTEM COMPONENTS

Computer-aided drafting systems consist of two basic elements, hardware and software. *Hardware* may be defined as the physical components of the system— electrical, electronic, electromechanical, magnetic, or mechanical devices. *Software* consists of sets of procedures, programs, and related documentation that directs operation of the system to produce graphics and related text material.

Hardware for a basic CAD system always consists of a computer that includes a central processing unit (CPU), a memory device, and a storage facility for programs and files. A typical CAD work station may also consist of a digitizer and puck, a graphics tablet and stylus, a function keyboard, an alphanumeric keyboard, a graphics display, and a plotter for producing graphical output. The system illustrated in Figure 10.2 is a complete interactive computer system for two-dimensional drafting.

10.4 MINICOMPUTER BASED SYSTEMS

The CADAM (Computer-Graphics Augmented Design and Manufacturing) System is an interactive graphics software system for computer-aided design and man-

ufacturing. Initially developed by the Lockheed Corporation, CADAM, Inc. is a wholly-owned subsidiary of the parent company, which continues to be one of its largest users. CADAM is designed to operate on IBM hardware of various sizes and configurations; from large mainframes to microcomputers in price ranges to accommodate most educational institutions' budgets. CADAM's high function, general purpose design and drafting system uses the IBM 3250 and 5080 Graphics Display Work Stations (GDWS). The IBM 3250 GDWS is shown in Figure 10.3.

The CAD (Computer-aided design) portion of the CADAM system uses a central design data base for creating two- and three-dimensional shapes using techniques similar to conventional drafting. Within the system are routines for developing oblique, isometric, and perspective views. In addition, CADAM has the ability to create a true three-dimensional data base for surface and wire frame geometry.

10.4.1 IMPORTANT ASPECTS OF THE SYSTEM

- CADAM has one of the fastest response times for operator actions of any existing system. The basic design criterion is to respond to a normal operator action within half a second. This high response time is a result of software design, in combination with the host computer power and specific display technology. Most users attribute the system's high productivity to the rapid response time.
- The CADAM system is highly interactive and user-oriented. Procedures at the graphics displays (IBM 3250s or 5080s) use construction techniques familiar to those of the conventionally trained drafter.
- Geometric construction is based on descriptive geometry. The graphics display is treated like a draft-

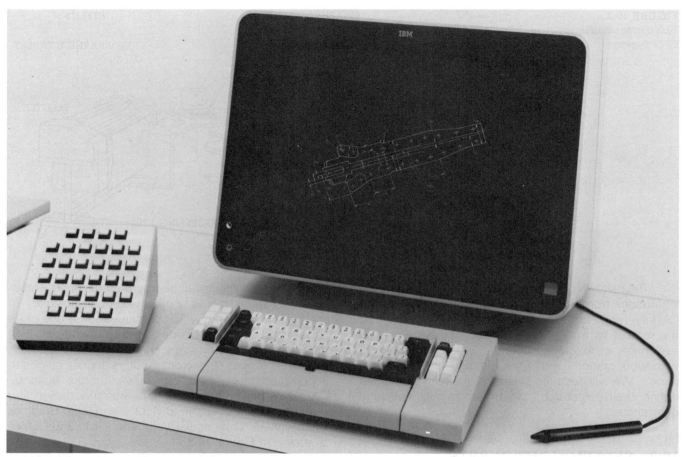

Courtesy of International Business Machines Corporation

FIGURE 10.3
IBM 3250 Graphics Display Work Station (GDWS).

ing board 20,000 inches × 20,000 inches in size. The user can zoom in and zoom out to get the desired resolution.

- In the area of system data compatibility and interference visualization, views can be merged, stacked, rotated, and juxtaposed.
- Replication of details is possible. A user may construct a detail, such as a fastener, bracket, or an electronics symbol only once, then replicate and locate it as often as necessary.
- Transformations help the drafter develop oblique and isometric views.
- 3-D surface geometry capability aids the operator in the design and visualization of 3-D surfaces. A user may construct ruled surfaces, bicubic surfaces, surfaces of revolution, and 3-D splines.
- With the aid of 3-D mesh generation facilities, designers may construct finite element models for structural analysis, heat-transfer analysis, and similar applications.
- A library of standard symbols can be stored in the system and called up as needed.

- User-defined special symbols, such as electronic configurations or architectural symbols, can be created and stored in special symbol tables, recalled for placement at different points, and scaled and rotated individually when placed.
- User-defined special character fonts (lettering styles) can be created and stored in special font tables to allow keyboard entry of characters in a variety of fonts with characters formed in any size, angle, and variable spacing.
- Hardcopy drawings can be obtained through plotter output. Microfilm output is also possible.
- English, metric, or dual dimensioning systems may be used.
- With the attribute facility, any part of an engineering drawing (single geometric element or permanent group) can be assigned user-defined attributes (weight, price, description, etc.). These data can be used as input to user programs to generate lists of materials and wiring instructions.
- Designers are able to generate better designs because they can test more design possibilities.

- Design changes made to components common to multiple subassemblies can be rapidly incorporated into many drawings.
- The data base is interdisciplinary, enabling an operator to make the geometry of a part available to other users.
- The data base system provides both data management and data protection. It controls the deletion of data and protects against unauthorized changes to drawings by means of a password-protect scheme.
- Expansion of a CADAM system is possible through the addition of more displays either at the central computer site or remotely on satellite processors.

10.4.2 APPLICATION AREAS

Because the CADAM system is a general-purpose computer-aided design and drafting system, it can be used anywhere there is a need to capture geometric data and work with it. These are some typical application areas for electrical and electronics drafting: layout for preliminary design; exploded views; perspective drawings; dimensioned drawings; isometric drawings; nondimensioned master drawings for production design; logic diagrams; charts, graphs, symbolic diagrams; tolerance between mating parts; volume, area, weight; machined parts; sheet metal parts; electrical systems; printed circuits; mechanical linkages; layouts and assemblies; schematic diagrams; connection and interconnection diagrams; and wiring harnesses and cabling.

10.4.3 GEOMETRIC CONSTRUCTION FACILITIES

The CADAM system has a number of geometric construction facilities that enable the user to define the geometry of a part and to construct a drawing. The operator interacts with the system through four input devices on the IBM 5080 or IBM 3250 graphics display—the alphanumeric keyboard, the 32-key program function keyboard, the light pen, or the tablet. All four devices are illustrated in Figure 10.4.

Courtesy of International Business Machines Corporation

Courtesy of International Business Machines Corporation

(a) ALPHANUMERIC KEYBOARD

(b) 32-KEY PROGRAM FUNCTION KEYBOARD

Courtesy of CADAM, Inc.

(c) LIGHT PEN
FIGURE 10.4
Input devices.

Courtesy of Summagraphics Corporation

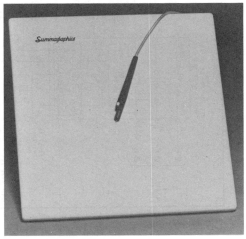

(d) GRAPHICS TABLET

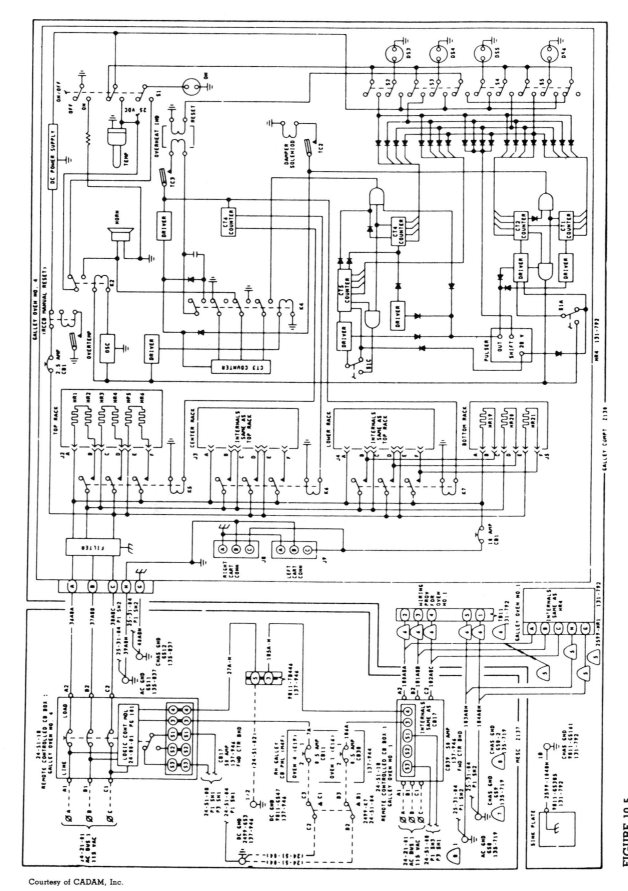

FIGURE 10.5
CADAM-produced electronics diagram.

108

Pressing a function key, for example, may cause the system to display a menu of geometric options to implement the function. The operator identifies the desired option by pointing the light pen at the option in the menu. The system acknowledges the user's choice and, if necessary, prompts the next step to be taken and the choice of available options. Thus, the operator might press the program function key to indicate that a line needs to be drawn. The system would offer several ways to define the line, such as:

- Horizontal
- Parallel (to some other line)
- Normal (to some other line)
- Angle (with some other line)
- Point-to-point

The operator selects an item by pointing at it with the light pen.

Besides a large number of similar geometrical figures, the CADAM system also has a number of display and data management functions to help the user construct a drawing—for example, zooming, translation, rotation, and copying.

This is a list of geometric construction, display management, and data management facilities available in the CADAM system:

Two-dimensional geometric construction facilities

a. Point
b. Line
c. Arc/Circle
d. Conics
e. Spline
f. Line Types
g. Composites
h. Trimming
i. Calculation and analysis functions

Three-dimensional geometric construction facilities

a. Surfaces
b. 3-D Splines
c. Intersection of plane and other surfaces
d. Network of nodes and connections

Two-dimensional projection of three-dimensional geometry

a. Display of management functions
b. Overlay
c. Auxiliary
d. Annotation
e. Data Management functions

10.4.4 CADAM SOFTWARE CONFIGURATIONS

The CADAM system is modular and versatile in construction and provides individualized user program (IUP) modules that can be combined to form various configurations.

- Interactive Design System: Accounting, CAD-GRAM, CADHUE, Data Management Hardcopy, Statistics, and Text Processing, Interface
- Manufacturing/Numerical Control: CADAMAC, APT Source Geometry, APT Interface, Compact II Interface, Split Interface
- Geometry Interface V.II: 3-D—3-D Interactive, 3-D MESH, 3-D Piping, D/B/M (Design/Build/Manage), and D/B/M Standard Library
  IPC—CADEX, CADPC, PRANCE Basic, PRANCE Multilayer, PRANCE ECL

An example of the type of electronics diagram that can be produced quickly and efficiently on CADAM is shown in Figure 10.5.

10.5 HP-EGS

Another two-dimensional CAD system is shown in Figure 10.6. It is a Hewlett-Packard Engineering Graphics System (HP-EGS) and is actually a computer-aided artwork tool to assist technical and professional staff in preparing engineering documents. The system consists of the primary configuration, which includes a desk-top computer, 1.6 megabytes of memory, on-line mass storage, a graphics tablet, disk drive, a Pascal 3.0 language system, and a plotter or printer.

FIGURE 10.6
HP-EGS—Engineering Graphics System.

10.5.1. SOFTWARE

The HP-EGS system has four basic personalities for producing computer-aided graphics:

- General Drawing—intended for general purpose drawing and as the foundation of a user-customized personality.
- Mechanical Engineering—mechanical drawing including dimensions, hatching, constructive geometry, and isometric views.
- Electrical Schematic Drawing—includes a library of commonly used parts to create schematics, along with the data required to create material and connection listings.
- Printed Circuit Board Layout—preparing printed circuit board artwork. It also contains the data necessary for material and connection listings.

A tablet menu is provided for the Mechanical Engineering, Schematic, and Printed Circuit Layout personalities.

The Graphics Editor software is the heart of this two-dimensional system, as indicated in Figure 10.7, and is used to generate drawings on the screen. It makes use of five types of files—message, menu, process, control, and macro, to define the user interface to the system. The engineering graphics system also contains two types of menus, a screen menu and a graphics tablet menu. Screen menus contain commands, library part names, and any other text needed to create drawing commands. Tablet menus consist of graphic images as well as textual messages.

Process files are used to describe the drawing layers, and 255 layers can be used on this system. For

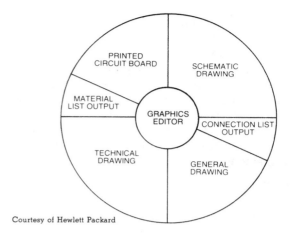

Courtesy of Hewlett Packard

FIGURE 10.7
Graphics editor.

each layer, the process file contains a layer label, a layer type, a linetype, color used when displaying the drawing, and the pen number to use when a plot is made. Eight different line types can be used, and as many as sixteen different colors can be used at one time with a color monitor. Units of measure on the system include micrometers, millimeters, centimeters, meters, kilometers, micro-inches, mils, inches, feet, yards, or miles.

10.5.2 DISPLAY

Figure 10.8 depicts the screen for the schematic drawing personality. Three major screen divisions can be seen in the figure: the main viewing area, the second-

FIGURE 10.8
Schematic drawing personality.

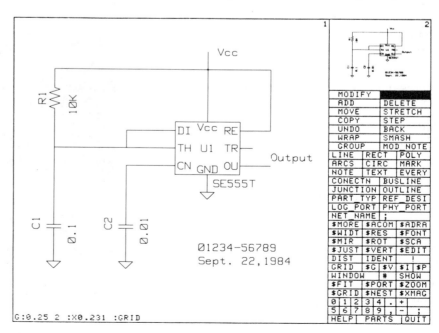

Courtesy of Hewlett Packard

FIGURE 10.9
HP-EGS keyboard.

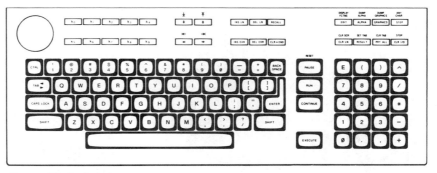

Courtesy of Hewlett Packard

ary viewing area, and the menu area. The cursor, which is controlled by the stylus on the graphics tablet, can access any part of the screen. Menu items are also selected by means of the stylus.

The operator interacts with the drawing through the primary (main) and secondary viewing areas of the screen. Commands allow for zooming in and out of the viewing area and also for panning across the screen. Commands operate in or between either the primary or secondary areas. For example, the secondary area can show a global view of the drawing while the primary area zooms in on a particular location. As modifications are made, the secondary area shows their relationship to the overall drawing. The cursor coordinates, expressed in the units specified, constantly change as the cursor is moved around the viewing area. The accuracy for drawing is .0001 inch with extremely fine grid resolution. All system commands can be entered from the keyboard, shown in Figure 10.9, as well as from the screen and tablet. The ten primitive drawing elements within the graphics editor are shown in Figure 10.10.

Each primitive element can be specified in several different ways. For example, when adding a circle to a drawing, the specific width of line may be specified along with whether or not the circle is to be filled. The graphics system also contains three different parts libraries that the operator can expand or change to fit specific applications needs.

10.5.3 THE PLOTTER

Production of a hard copy of the drawing occurs at the plotter. Three different-size plotters are available for this system, each of which has its own special features. The plotters will plot on single sheets from "A" (8½ inches × 11 inches) to roll sizes depending on the model of plotter. Plotting is quick and precise, and lines and curves sharply defined. *Resolution* on a plotter can be defined as the smallest move that can be specified programmatically. The resolution on each model is 0.025mm (0.000984 inches). High repeatability allows the plotter to join new lines smoothly to previously plotted ones. The drafting plotter holds eight pens of three types: fiber tip, roller ball, and drafting. Each pen type is mounted in a carousel as pictured in Figure 10.11. When not in use, pens remain tightly capped in the carousel to prevent premature ink drying. Four buttons control normal plotter operation, as shown in Figure 10.12. These buttons are used for medium (paper, mylar) loading, and tell the plotter to begin accepting program instructions from the computer. They allow interruption of plotting so the operator can view the drawing as it is being produced. The front panel also provides local control of many plotter functions. (The panel is illustrated in Figure 10.13.) For example, the operator can override default or programmed values for pen speed and force, reset plotting limits, change P1 and P2 positions for scaling, rotate the drawing 90

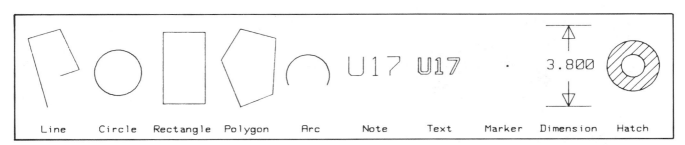

Courtesy of Hewlett Packard

FIGURE 10.10
Primitive drawing elements.

FIGURE 10.11
Pen carousel.

FIGURE 10.12
Normal operations—Plotter control panel.

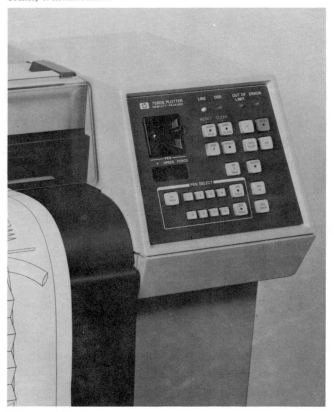

FIGURE 10.13
Full control panel.

degrees, and align the axes with those on preprinted forms.

10.6 MICROCOMPUTER BASED SYSTEMS

Several microcomputer-based CAD systems are available in the marketplace. They can perform simulation, analysis, drafting, and design functions. Several software programs are available on the more powerful systems and more programs are becoming available almost daily.

10.6.1 THE CADAPPLE 2-D DRAFTING SYSTEM

This system was developed for an APPLE II or APPLE IIe computer with the following minimum hardware configuration: computer with 64K bytes of memory, keyboard, monitor, two standard 5¼-inch floppy disk drives (single side, single density), joystick, and plotter. This configuration is shown in Figure 10.14.

As with all computers, the central processor is the brain of the system. With the keyboard and joystick or digitizer (input device), one enters information into

FIGURE 10.14
Apple IIe personal computer.

FIGURE 10.15
5¼-inch floppy disk.

the computer. After input, the information is stored on a removable recording medium called a *floppy disk* or *diskette.* A floppy disk is a thin piece of mylar encased in a protective plastic jacket with a surface that can be magnetized to store data. Figure 10.15 shows an example of a 5¼-inch floppy disk. The plotter (output device) is used to provide a copy of the drawing that is entered into the computer. The display monitor (screen) displays both text and graphics. The CADAPPLE system provides these convenient features:

- The program is "menu driven."
- The program indicates to the operator what options are available.
- The operator sees on the screen exactly what the final plot will look like before it is plotted.
- The program is interactive; that is, the operator receives immediate visual and audio feedback on all actions.
- The program has predefined "default" actions and values that enable the novice user, with only a minimal understanding of the program's operation, to produce usable drawings.
- If the user finds that an unknown path has been taken, he may press the escape (ESC) key on the alpha keyboard to abort the erroneous action.

10.6.1.1 Input Terminology
The 2-D program receives its graphics input from the digitizer (sometimes called the graphics tablet) or from the joystick. The cursor's position on the input device is represented on the screen by a small crosshairs. As the drafter/designer moves the input device cursor, the crosshairs move across the screen. With the cursor, one can create various graphic objects such as lines,

circles, and rectangles and can graphically manipulate the entire picture or any part thereof.

10.6.1.2 The Menu
This program is organized around a collection of many "menus." As with all menus, each menu has a specific purpose. For example, the FILER menu allows entry of commands for manipulating the graphics files, and the OUTPUT menu allows the operator to send the graphics drawing to an output device such as a plotter.

10.6.1.3 Objects
As is true with other programs on other systems, this software allows the user to create a drawing using basic building blocks called objects: basic line shapes, dimensions, points, and text.

10.6.1.4 Levels
The CADAPPLE program allows the user to manipulate up to 250 different levels, also referred to as *layers* or *sheets*, on systems previously illustrated.

10.6.1.5 Group
The program also allows the user to manipulate objects in the form of "groups." A group is an arbitrary user-defined collection of objects. The only relationship that objects in a group have to one another is that they share the same group name. Groups provide a convenient way of manipulating large components of a complex drawing. They allow the operator to move, copy, and otherwise manipulate large numbers of objects at one time.

10.6.1.6 Windows
The user may view the drawing through any user-defined viewing *window,* also called a *zoom-in, zoom-out,* or *pan* feature on other systems.

10.6.1.7 Properties

As primitive objects are created, they are given several attributes or properties that partly determine what they will look like. An example of a property is the type of linestyle (dashed or dotted) that will be used to plot the object. The user can change these default property values at any time.

10.6.1.8 Real World Coordinates

All numeric input to the 2-D program is in terms of real-world X, Y coordinates. The coordinates reflect the actual size of the object being drawn; for example, if an architect were designing a room addition for a house, all the information about the room would be entered at actual scale, that is, if the width of the room is 20 feet, then the lines making up the walls of the room would be entered as 20 feet. All scaling is done when a plotted hardcopy is needed.

10.6.2 THE AUTOCAD SYSTEM

AutoCAD is another two-dimensional computer-aided drafting and design system that operates on low-cost microcomputers. It is produced by Autodesk, Inc. and is a multi-utility system, suitable for applications in electronics and other technical drafting areas. It follows instructions to help produce a drawing quickly, offers features that allow for correction of drawing errors, and makes even large revisions without redoing an entire drawing. It produces clean, precise, final artwork. It puts nothing into the artwork "on its own." A completed AutoCAD drawing looks identical to the same drawing carefully prepared by hand, except that it is probably more accurate. The drawing is configured exactly as specified, with every element appearing where it is wanted.

10.6.2.1 Equipment Requirements

The AutoCAD system is a very flexible software package and currently operates on the following hardware: Victor 9000; IBM PC; IBM PC/AT; IMB PC/XT; IBM 3270 PC/G; IBM 3270 GX; Fujitsu MIG; Tandy 2000; Wang PC; DG/ONE; Apricot; HP 150 Touch Screen; Zenith Z100; NEC APC; Columbia; Texas Instruments Professional; NCR PC; DMS; and DEC Rainbow. A pictorial of an IBM PC, PC/AT and PC/XT microcomputer and a Digital Equipment Corporation Rainbow 100 microcomputer are shown in Figures 10.16, and 10.17. AutoCAD supports a variety of input and output devices, including Sun-Flex, Houston Instruments, Summagraphics, Hitachi, GTCO, USI Opto-Mouse, Logitech, Microsoft, Mouse Systems, CalComp, SCION, Gould, Hewlett Packard, IBM, Roland, Strobe, Sweet-P, Western Graphtec, Tecmar, Hercules, and Vectrix. In addition to a basic computer system (including processor, keyboard, text display screen, and disk

drives), AutoCAD requires a graphics monitor capable of reasonably high resolution.

10.6.2.2 General Operation

A graphics monitor is used to display the drawing. Everything done to the drawing appears on the monitor so the user can watch each step of the progress as on other systems. Sometimes two monitors are preferable so the operator can view the menu on one while inputting graphics on the other.

AutoCAD provides a set of entities for use in constructing the drawing. An *entity* is a drawing element also referred to as *objects* or *primitives* on the systems we have previously discussed. Commands are entered to tell AutoCAD which entity to draw. Commands may be typed on the keyboard, selected from a screen menu using a pointing device, or entered with the push of a button from a menu on a digitizing tablet. Then, following prompts on the display screen, certain information is specified for the chosen entity. This always includes the point in the drawing where the operator wants the entity to appear; sometimes a size, angle, or rotation value is also required. After this information is supplied, the entity is drawn and appears on the graphics monitor. A new command may then be entered to draw another entity or perform a different AutoCAD function.

The program provides commands that allow modification of the drawing; entities can be erased, moved, or copied to form repeated patterns. Other commands let the user change the view of the drawing displayed on the graphics monitor, provide special information about the drawing, and offer drawing aids that help position the entities accurately. Each action appears immediately on the graphics monitor. Plotting can be accomplished with a simple command.

10.6.2.3 Terminology for AutoCAD

- AutoCAD Drawing—An AutoCAD drawing is a file of information that describes a graphic image. It may be any size, specified by any unit of measurement, and corresponds exactly to a drawing prepared on paper. That is, "entities" in the drawing (elements such as lines, circles, text, etc.) are positioned in the drawing file exactly where they would be on paper.
- Coordinates—A coordinate system is used to locate points on the drawing; for example, to position entities. An X coordinate specifies vertical location. Thus any point on the drawing can be indicated by an X and Y coordinate pair.
- Drawing units—As noted, entities in the drawing are positioned on coordinate points; for example, a line is drawn by specifying the coordinates of its two end points. The distance between two points is measured in units. Thus, a line drawn between the

points (1,1) and (1,2) is one unit in length. A unit corresponds to whatever form of measurement the drawing requires, whether inches, feet, centimeters, angstroms, or other incremental units. When

Courtesy of International Business Machines Corporation

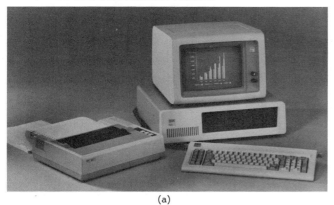

(a)

Courtesy of International Business Machines Corporation

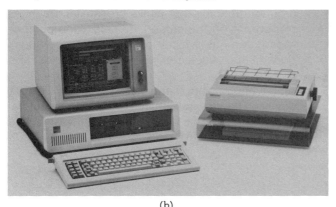

(b)

Courtesy of International Business Machines Corporation

(c)

FIGURE 10.16
Microcomputer systems: (a) IBM PC; (b) IBM PC/XT; (c) IBM PC/AT.

the drawing is plotted, the user may specify a scale factor to plot each unit the exact required size.

■ Display—The term *display* is used in two ways. First, it indicates the graphics monitor screen upon which the picture is shown. More often in AutoCAD, however, display refers to that portion of the drawing that is currently being displayed. The display may be zoomed in or out to magnify or shrink the visible image, called *windowing* on the CADAPPLE system.

■ Resolution—Physical resolution refers to the amount of detail that can be represented. For digitizing tablets and plotters, resolution is usually specified as "dots per inch." Digitizer resolution determines the precision with which the operator can indicate closely-spaced points. Plotter resolution determines the "smoothness" and precision of the plotted artwork. Entered coordinates can be aligned to the nearest point on an optionally visible grid. Spacing of the grid points is called *snap resolution*, resolution of the coordinates in the drawing data base, which is completely independent of the resolution of input and output devices. In printed circuit drafting, for example, it is common to make all points align on 0.1-inch centers. In that case, AutoCAD would be set to a snap resolution of 0.1. Snap resolution may be set to any "fineness" the artwork requires.

10.6.2.4   AutoCAD Features

■ Main menu—AutoCAD operates on two levels to reduce both the work necessary for generating a drawing and the time needed to learn the system. At the outer level, AutoCAD provides a menu-driven interface (the "main menu") that allows the user to

Courtesy of Digital Equipment Corporation

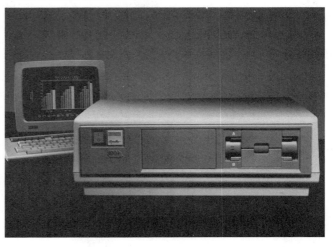

FIGURE 10.17
Digital Equipment Corporation (DEC) Rainbow 100.

initiate various tasks, such as creating new drawings, modifying stored drawings, producing plots, and so on. At the inner level, the menu provides access to various parts of AutoCAD, such as the Interactive Drawing Editor and the plotter interface.

■ Interactive drawing editor—The drawing editor displays the intended picture and provides commands to create, modify, view, and plot drawings. AutoCAD automatically loads the drawing editor when certain tasks are selected from the main menu.

■ Screen menu—A menu can be displayed on the graphics monitor while the drawing editor is active. This menu allows command entry simply by pointing to the command on the display screen.

■ Drawing insertion—This feature allows the operator to insert an existing AutoCAD drawing (stored on disk) into the drawing currently being created. Thus, the user can interactively construct a drawing "part," store it in a regular AutoCAD drawing file, then insert as many copies as necessary in subsequent drawings.

■ Layers and colors—Parts of the drawing may be assigned to any of 127 different layers. (We have already discussed layers and overlays on other systems.) A color may be associated with each layer. The color is simply a number from 1 to 127; what it means depends on the output device. For monochrome devices, it has no effect; for multipen plot-

ters, however, a particular pen and line type can be associated with each color number.

■ Graphic input (pointing) devices—A pointing device may be used for quick command entry and to locate points in the drawing. Several types of pointers can be used with AutoCAD as well as with similar software packages for microcomputers. They include:

a. Light Pen—With a light pen, one can point directly at an area of the graphics monitor's screen to enter points or select commands from a "screen menu." Crosshairs appear on the screen and follow the pen until a point or menu item is selected by releasing or deactivating the pen.

b. Touch Pen—This device operates much like a light pen, but works by touching a special panel on the face of the graphics monitor. One selects a point or menu item by sliding the pen tip along the panel until the crosshairs are positioned at the desired point, then lifting the pen tip from the screen.

c. Mouse—A *mouse* is operated by moving it around the tabletop while crosshairs track its movement on the screen. To select the point or menu item at which the crosshairs are positioned, the user pushes the button on the mouse.

d. Tablet—Point and menu item selection using a digitizing tablet are similar to the mouse operation; however, the tablet's stylus moves around

Courtesy of Summagraphics Corporation

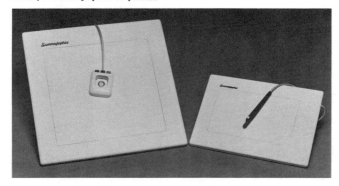

Courtesy of Summagraphics Corporation

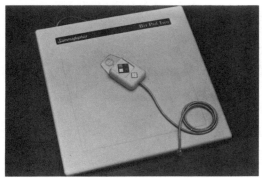

Courtesy of Summagraphics Corporation

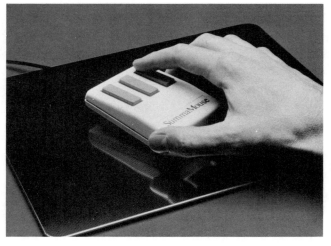

FIGURE 10.18
Graphic-input (pointing) devices.

only on the tablet's surface. Figure 10.18 shows examples of graphic input (pointing) devices.

- Point and command entry flexibility—One can enter commands and specify points in the drawing in a variety of ways. From the keyboard, points can be specified by typing in absolute coordinates or coordinates relative to the last point specified; or by keyboard pointing, which utilizes keyboard control keys to move crosshairs around on the graphics monitor so the user can visually position to the desired point.
- Tablet menu—A menu of AutoCAD commands can be placed on the digitizing tablet, permitting entry of a command simply by pointing to it with the stylus and pushing a button.

- Database storage—All information about the drawing, size and position of every element, the size of the drawing itself, its display characteristics (e.g., whether "zoomed" in or out), and so forth is automatically updated with each command and stored on disk.
- Plotting—A hard copy of the drawing can be plotted at any stage of its development. "Check-plots" can be generated while the drawing is in progress to check for positioning and dimensioning errors that might not be immediately apparent on the screen. (Typical hard-copy plotters for microcomputer based systems are shown in Figure 10.19.)

A package of advance features is available for the system as extensions to the basic AutoCAD software.

9872 C Plotter, courtesy of Hewlett Packard

7470 A Plotter, courtesy of Hewlett Packard

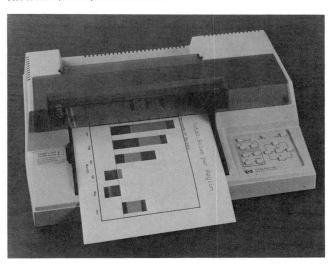

DMP-40 Plotter, courtesy of Houston Instrument

DMP-29 Plotter, courtesy of Houston Instrument

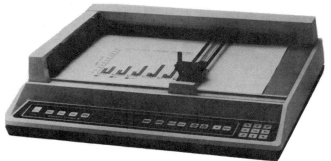

FIGURE 10.19
Hard copy plotters for microcomputer systems.

10.7 SUMMARY

Computer-aided drafting (CAD) has had a tremendous positive impact on the way business is conducted in engineering departments, particularly in the design and drafting functions. One of the greatest advances has been in the field of technical graphics, especially in the production of electronic and electrical diagrams and drawings. Productivity has increased, and at the same time, cleaner, neater, and more accurate drawings are being produced.

Equipment ranging from stand-alone mainframe systems to minicomputer and microcomputer hardware and software is available. We have examined several mini and micro two-dimensional drafting systems. These systems are generally within the financial reach of most educational institutions, which can acquire the equipment for student and staff development, instruction, and training. We discussed in detail representative examples of CAD software packages. New software packages are rapidly being developed so that additional hardware and software capabilities continue to proliferate. Section 10 Practice Exercises are on pages 297 through 303.

10.8 BASIC CAD TERMINOLOGY

Alpha characters The 26 letters of the alphabet (A through Z).

Alphanumeric keyboard A keyboard similar to the typewriter keyboard. Allows the user to input letter (alpha) and number (numeric) instructions to the central processor.

Backup disk A disk containing a copy of another disk; created for protection in case one disk becomes damaged or lost.

Binary The base 2 numbering system; uses only digits 0 and 1.

Bit Taken from binary digit; 0 or 1 signal.

Byte Eight binary digits (bits) that the computer operates on as a single unit. Used as the basis of comparison in describing various systems and manufacturers. One byte is a character of memory; a *megabyte* is 1 million characters.

Command An instruction to the system to perform a function.

Computer Popular name referring to the CPU.

CPU Central processing unit; the microprocessor portion of the computer that accomplishes the logical processing of data. The CPU contains the arithmetic, logic, and control circuits, and possibly the memory storage.

CRT A cathode-ray tube that projects electrons onto a fluorescent screen to produce graphic displays on video screens.

Cursor A moving bright marker on the screen showing the placement of the next character; may be variously shaped as a dot, crosshair, check, or rectangle.

Digitizer An electronic tracing board used to enter existing drawings into the system by means of a pointer and a menu.

Digitizing "Touching down" a stylus or puck at a particular location, the method by which data is entered on a graphics tablet.

Disk drive A device that holds, reads, and writes onto disks. Most systems use two disk drives.

Display The screen on which a drawing is displayed; a method of representing information in visible form. Display screens are available in a monochrome (black and white) or color format.

Down time Any interval of time when the system is not available or not working.

Enter To put information into the system.

Execute To carry out a command or a series of commands.

File A collection of information treated as a unit, with an identifying name assigned to it.

Floppy Disk A portable, flat, black, circular plate made of thin vinyl with a magnetic coating that is inserted into and removed from the floppy disk drive. It is a software package for storing data, available in standard sizes of 5¼ inches and 8 inches in diameter.

Font A set of alphanumeric characters and symbols drawn in a particular typeface.

Function A type of calculation performed by the system; examples are *geometric functions* and *text functions.*

Graphic display The unit that displays the image or drawing. The most popular graphic display is the CRT (cathode-ray tube).

Graphics tablet An input device having a flat surface on which the work is done. A stylus or a puck is used for the graphical data entry and information is transmitted to the CRT by means of an electrically controlled grid beneath the tablet's surface.

Hard copy A preliminary drawing produced by the hard-copy unit and often used as a checkprint.

Hard disk A nonremovable disk with a large capacity for storing information in the form of files; sometimes referred to as a *Winchester.*

Hardware The physical equipment necessary for system operation.

Host One central place where the data resides.

Joystick An input device that directly controls the cursor. One moves the stick in the same direction that one wishes the cursor to move on the screen.

Light pen An input pointing device. Data entry may be made directly onto the screen by positioning and activating the tip of the pen at the desired location.

Mainframe A CPU with a large capacity and many terminals for multipurpose use.

Memory Stored information, programs, and data inside automated equipment. One byte is a character of memory.

Menu An area of the digitizer containing preassigned commands that control system operations.

Menu tablet An input device having a flat surface on which functions are selected with a stylus or puck; also known as a *menu pad.*

Message Any words or sentences displayed on the screen that inform the user about commands being executed or information needed by the system.

Microcomputer The smallest type of CAD system; micros are dedicated units using personal computers.

Minicomputer A CAD system with capabilities ranging between those of a micro and a mainframe. Because of their size, they are often referred to as desktop computers. This type of CAD system is commonly used in industry.

Modem An electronic device that allows computer equipment to send and receive information through telephone lines.

Monitor The unit that displays the image or drawing; same as CRT, graphics display, or screen.

Peripheral Additional equipment working in conjunction with, but not as part of, the computer.

Pan To move the display of a drawing on the screen, much as a camera moves to find the best picture.

Plot The final drawing as it is drawn by the plotter on paper, vellum, or polyester film.

Plotter The piece of equipment that draws mechanically, on paper, vellum, or polyester film, the drawing from the screen display.

Prompt The arrow ()), asterisk (*), or dot (·) sign at the beginning of a line on the screen signifying that the system is ready for one to enter a command.

Puck A manually operated control device used to input data.

Raster A network or matrix of dots; each dot falls within a square area known as a *pixel.*

Resolution The number of addressable dots per unit area. Low-resolution screens produce jagged, stepped lines.

Remote station Another work station or system connected to the local system with a modem to enable exchange of information over telephone lines.

Screen The display on a video monitor on which drawings or messages are viewed.

Software The collection of programs used to control the system's internal functioning and hardware.

Stylus A manually operated device that provides input to the display unit.

Terminal Popular name for the combination of a visual display screen (CRT) and keyboard, sometimes referred to as a video display terminal (VDT).

Turnkey A system that operates with a variety of software and whose hardware is functional, complete, and ready to use as soon as installation is complete (ready at the turn of a key).

Winchester disk A type of hard disk that is nonremovable.

10.9 REVIEW EXERCISES

1. Why is a CAD system valuable?

2. Lay out freehand a block diagram of the basic concept of a CAD system.

3. Name six types of electronic diagrams and drawings that can be easily produced on a CAD system.

4. Define CAD "hardware."

5. Define CAD "software."

6. What does the term "interactive" signify?

7. What three types of computer systems are available for CAD work?

8. What is a "menu tablet"?

9. What piece of hardware in a CAD system produces a "hard copy"?

10. What is the purpose of an alphanumeric keyboard?

11. What is a "floppy disk"?

12. What is the function of a cursor?

13. What is a "CPU"?

14. What is an "input device"? Give an example.

15. What is an "output device"? Give an example.

ACOUSTIC DEVICES

    HORN, ELECTRICAL; HOWLER, LOUDSPEAKER, SIREN

        GENERAL

        LOUDSPEAKER - MICROPHONE

    MICROPHONE, TELEPHONE TRANSMITTER

    TELEPHONE RECEIVER, EARPHONE, HEARING-AID RECEIVER.

    HANDSET, OPERATOR'S SET

        GENERAL

BATTERY

    The long line is always positive, but polarity may be indicated when required.

    ONE CELL

    MULTICELL

CAPACITOR

    When it is necessary to identify the capacitor electrodes, the curved element
    shall represent the outside electrode in fixed paper and ceramic capacitors,
    the moving element in adjustable and variable capacitors and the low
    potential element in feed-through-capacitors.

    GENERAL (FIXED)

    POLARIZED

    SHIELDED

VARIABLE

FEED-THROUGH (With terminals shown on feed-through element for clarity)

CIRCUIT PROTECTORS
  FUSE

      GENERAL

OR

COILS
  GENERAL, PREFERRED SYMBOL

      ALTERNATE SYMBOL

  CHOKE, RADIO FREQUENCY 
  INDUCTANCE
    FIXED

      ADJUSTABLE

  ADJUSTABLE, FERRITE

  ADJUSTABLE OR CONTINUOUSLY VARIABLE

  INDUCTION COIL
    TWO WINDING

      THREE WINDING

  REPEATING

  TRANSFORMER

      MAGNETIC CORE TRANSFORMER

CONNECTOR
  DISCONNECTING DEVICE

    CONTACTS

      The contact symbol is similar to an arrowhead except the lines are drawn at
      a 90°angle.

        A - MALE CONTACT, PLUG (One Per Line)

        B - FEMALE CONTACT, JACK (One Per Line)

        C - ENGAGED CONTACTS, MATING PAIR

    ASSEMBLY

        A - PLUG END (Usually Movable)

        B - RECEPTACLE (Usually Stationary)

        C - ENGAGED CONNECTORS

        D - APPLICATION:

          This engaged 4 conductor connector plug has 1 male and 3 female
          contacts.

        * Show appropriate contact designation (numbers and/or letters) and
          type of contact (male or female) applicable to connector configuration.

CONTACTS, RELAYS, SWITCHES

  SWITCHING FUNCTIONS

    CONDUCTING, CLOSED CONTACT (BREAK)

    NON-CONDUCTING, OPEN CONTACT (MAKE)

COMMONLY USED GRAPHIC SYMBOLS FOR ELECTRICAL AND ELECTRONIC DIAGRAMS	APPENDIX A

TRANSFER (BREAK-MAKE)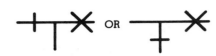

ELECTRICAL CONTACT
  FIXED CONTACT

A - FIXED CONTACT FOR JACK, KEY, RELAY, ETC.

B - FIXED CONTACT FOR SWITCH

C - FIXED CONTACT FOR MOMENTARY SWITCH

D - SLEEVE

  MOVING CONTACT

A - ADJUSTABLE OR SLIDING CONTACT FOR RESISTOR,
    INDUCTOR, ETC.

B - LOCKING CONTACT

C - NON-LOCKING CONTACT

D - SEGMENT OR BRIDGING CONTACT

E - VIBRATOR REED

F - ROTATING CONTACT (SLIP RING) AND BRUSH

BASIC CONTACT ASSEMBLIES
    The standard method of showing a contact is by a symbol showing
    the circuit condition it produces when the actuating device is in the
    de-energized or non-operated position.

OPEN CONTACT (MAKE)

CLOSED CONTACT (BREAK)

TRANSFER (BREAK-MAKE)

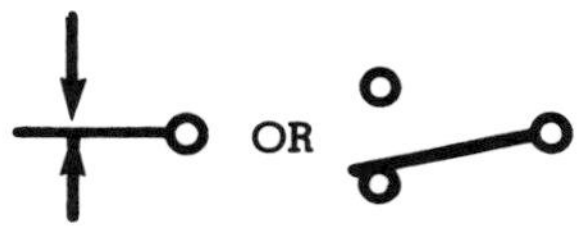

MAKE BEFORE BREAK

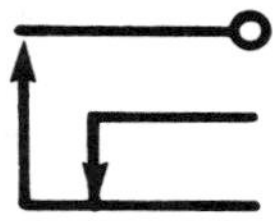

SWITCH, GENERAL

    SINGLE POLE SINGLE THROW (SPST)

    SINGLE POLE DOUBLE THROW (SPDT)

    DOUBLE POLE SINGLE THROW (DPST)

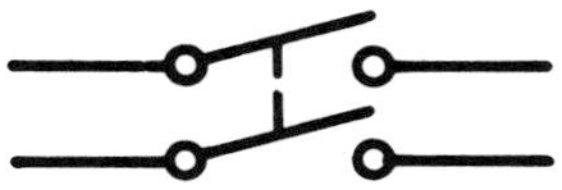

    DOUBLE POLE DOUBLE THROW (DPDT)

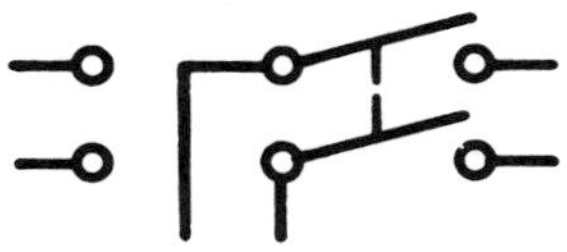

    TRIPLE POLE SINGLE THROW (TPST)

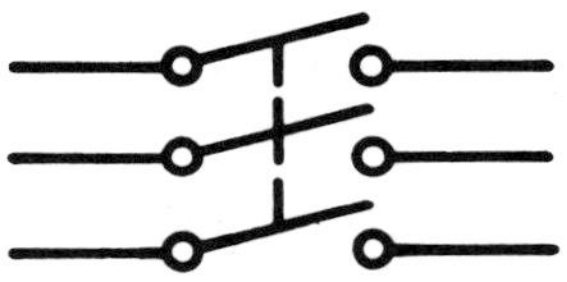

    TRIPLE POLE DOUBLE THROW (TPDT)

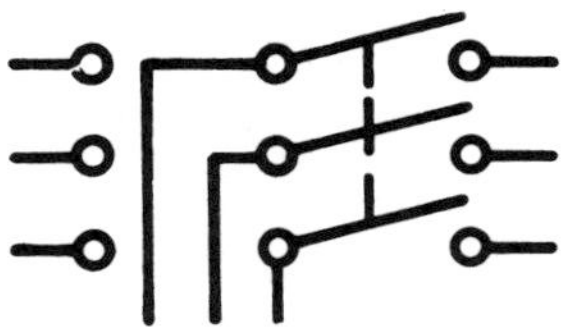

PUSHBUTTON SWITCH, MOMENTARY OR SPRING RETURN

    CIRCUIT CLOSING (MAKE)

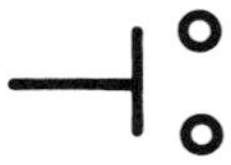

    CIRCUIT OPENING (BREAK)

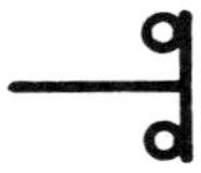

COMMONLY USED GRAPHIC SYMBOLS FOR ELECTRICAL AND ELECTRONIC DIAGRAMS	APPENDIX A

SELECTOR OR MULTIPOSITION SWITCH

The position in which the switch is shown may be indicated by note or designation of switch position.

BREAK-BEFORE-MAKE. NONSHORTING (NON-BRIDGING) DURING CONTACT TRANSFER.

MAKE-BEFORE-BREAK. SHORTING (BRIDGING) DURING CONTACT TRANSFER.

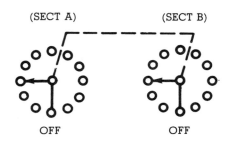

ROTARY SWITCH, WAFER TYPE. TWO SECTION, CONTACTS ON ONE SIDE. (NONSHORTING)

ROTARY SWITCH, WAFER TYPE. ONE SECTION, CONTACTS ON BOTH SIDES. (SHORTING)

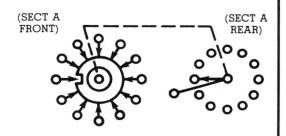

DELAY LINE

GENERAL

TAPPED DELAY FUNCTION

Note: The two vertical lines indicate the input side.

* Length of delay may be indicated.

ELECTRON TUBES

    VACUUM TUBES

        MULTI-GRID VACUUM TUBE

        SINGLE THIODE VACUUM TUBE

        TWIN TRIODE VACUUM TUBE

           Note: Tube filaments may be shown detached from
               the tube symbol to show the filament
               connections more conveniently.

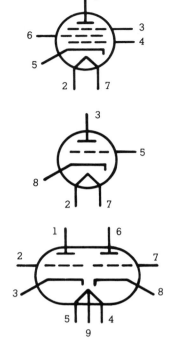

    GAS FILLED TUBES

        GRID CONTROLLED GAS TUBE (THYRATRON)
        Note: Black dot signifies gas filled.

        COLD CATHODE GAS TRIODE TUBE

        COLD CATHODE GAS TRIODE TUBE
        EQUIPPED WITH AN EXTERNALLY
        MOUNTED RESISTOR IN THE STARTER
        ANODE (RED) LEAD.

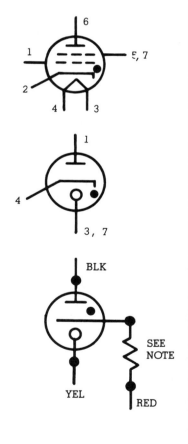

COMMONLY USED GRAPHIC SYMBOLS FOR ELECTRICAL AND ELECTRONIC DIAGRAMS	APPENDIX A

LAMPS AND VISUAL SIGNALING DEVICES
  LAMP

    BALLAST LAMP

    FLUORESCENT LAMP

      TWO TERMINAL

      FOUR TERMINAL

    GLOW LAMP, COLD-CATHODE LAMP, NEON LAMP
      ALTERNATING CURRENT TYPE

      DIRECT CURRENT TYPE

    INCANDESCENT MAZDA LAMP

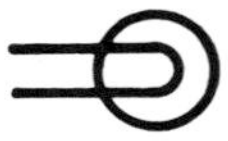

    INDICATING LAMP, SWITCHBOARD   ✻
      * To indicate the following characteristics the specified letter(s) may be 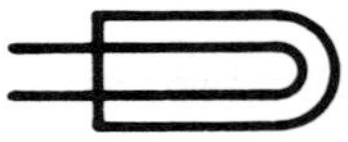
      placed within or adjacent to the symbol:

        A - Amber   OP - Opalescent
        B - Blue   P - Purple
        C - Clear   R - Red
        G - Green   W - White
        O - Orange   Y - Yellow

    JEWELED SIGNAL LIGHT   ✻

  METER, INSTRUMENT

    GENERAL   ✻ 
    * Add the following letter(s) to indicate the function of the meter.

        A - AMMETER   G - GALVANOMETER
        AH - AMPERE HOUR   MA - MILLIAMMETER
        CMC - CONTACT-MAKING   OHM - OHMMETER
        (OR BREAKING) CLOCK   UA - MICROAMMETER
        F - FREQUENCY METER   V - VOLTMETER

COMMONLY USED GRAPHIC SYMBOLS FOR ELECTRICAL AND ELECTRONIC DIAGRAMS	APPENDIX A

RESISTOR

    FIXED

    TAPPED

    ADJUSTABLE

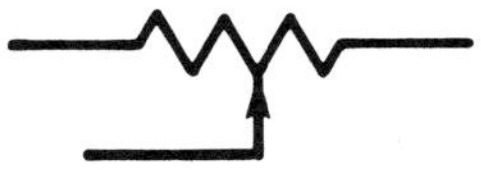

    ADJUSTABLE OR CONTINUOUSLY VARIABLE

    INSTRUMENT OR RELAY SHUNT (CONNECT INSTRUMENT OR RELAY TO
    TERMINALS IN THE RECTANGLE.)

    VARISTOR

    THERMISTOR

    THERMISTOR WITH INDEPENDENT INTEGRAL HEATER

SEMICONDUCTOR DEVICES

    TWO TERMINAL DEVICES

        RECTIFIER DIODE, METALLIC RECTIFIER

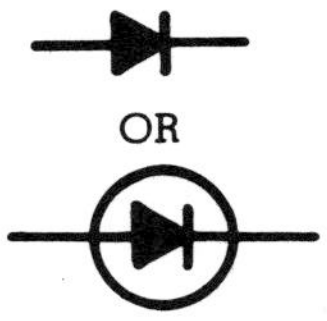

        OR

        CAPACITIVE DIODE (VARACTOR)

        TEMPERATURE DEPENDENT DIODE

        PHOTO SENSITIVE DIODE

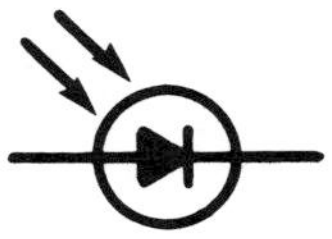

COMMONLY USED GRAPHIC SYMBOLS FOR ELECTRICAL AND ELECTRONIC DIAGRAMS	APPENDIX A

PHOTO EMISSIVE DIODE

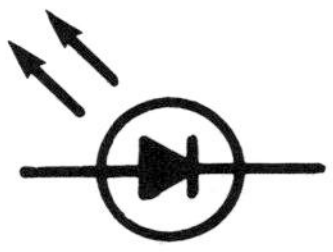

STORAGE DIODE

BREAKDOWN (ZENER) DIODE (UNIDIRECTIONAL)

BREAKDOWN DIODE (BIDIRECTIONAL)

UNIDIRECTIONAL NEGATIVE-RESISTANCE BREAKDOWN DIODE

   NPN TYPE

   PNP TYPE

BIDIRECTIONAL NEGATIVE-RESISTANCE BREAKDOWN DIODE

   NPN TYPE

   PNP TYPE

TUNNEL DIODE    OR

BACKWARD DIODE, TUNNEL RECTIFIER    OR

THYRISTOR, REVERSE-BLOCKING DIODE TYPE   OR
(ALSO FOUR LAYER OR STOCKLEY DIODE)

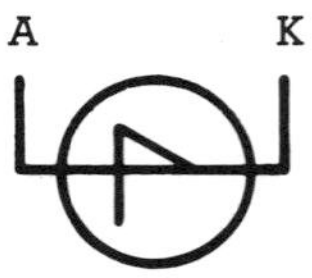

THREE (OR MORE) TERMINAL DEVICES

    PNP TRANSISTOR

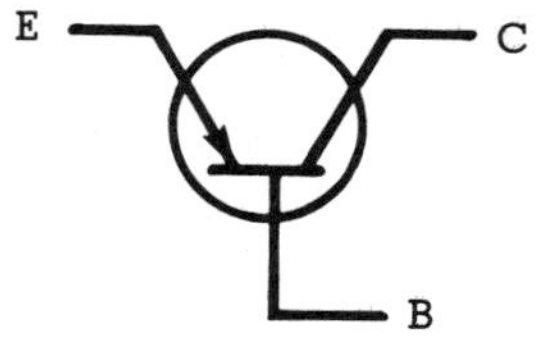

    NPN TRANSISTOR

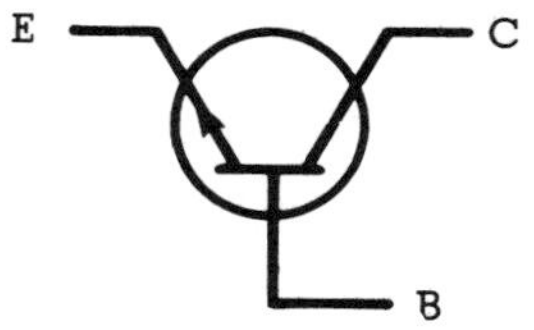

    THYRISTOR, SEMICONDUCTOR CONTROLLED RECTIFIER

    THYRISTOR, SEMICONDUCTOR CONTROLLED SWITCH

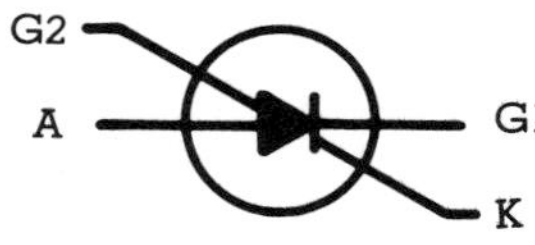

    UNIJUNCTION TRANSISTOR WITH N TYPE BASE

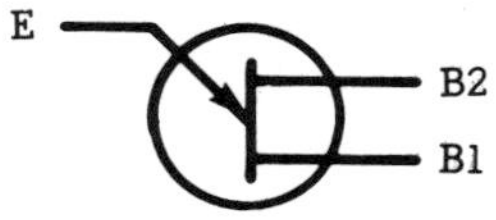

    UNIJUNCTION TRANSISTOR WITH P TYPE BASE

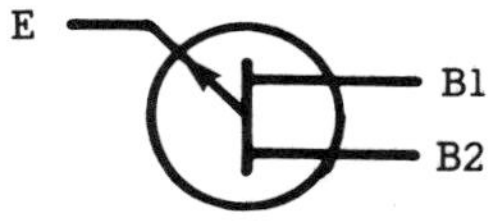

    BRIDGE TYPE RECTIFIER

WIRING CONVENTIONS

    CABLE, CONDUCTOR, WIRING

        CONDUCTIVE PATH OR CONDUCTOR WIRE

COMMONLY USED GRAPHIC SYMBOLS FOR ELECTRICAL AND ELECTRONIC DIAGRAMS	APPENDIX A

TWO CONDUCTORS OR CONDUCTIVE PATHS

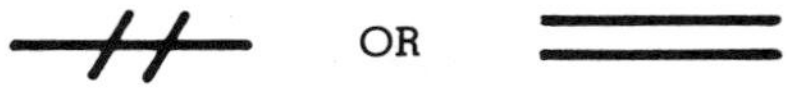

THREE CONDUCTORS OR CONDUCTIVE PATHS

"N" CONDUCTORS OR CONDUCTIVE PATHS
    Note: The actual number of conductors or conductor
        paths is substituted for the letter "N."

CROSSING OF CONDUCTORS OR CONDUCTIVE
PATHS, NO CONNECTION. (THE CROSSING NEED
NOT BE AT A 90° ANGLE.)

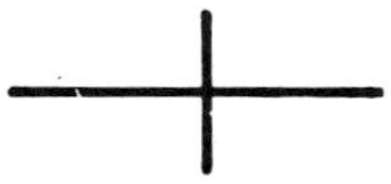

JUNCTION OF CONDUCTORS OR CONDUCTOR PATHS

$\frac{3}{32}$ DIA.

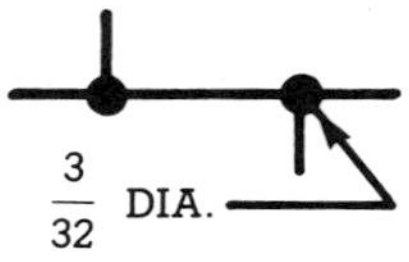

TWISTED PAIR

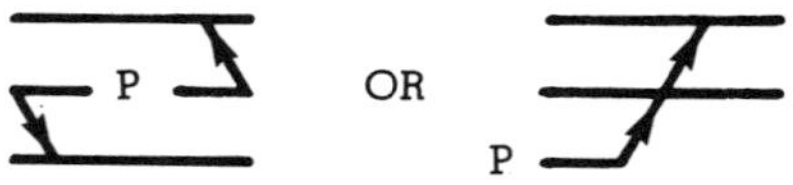

THREE WIRE TWISTED

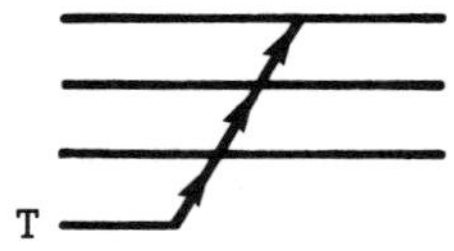

FOUR WIRE TWISTED

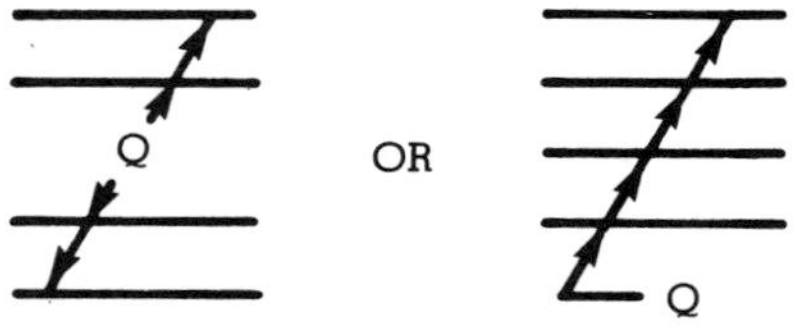

SHIELDED WIRE, SINGLE CONDUCTOR

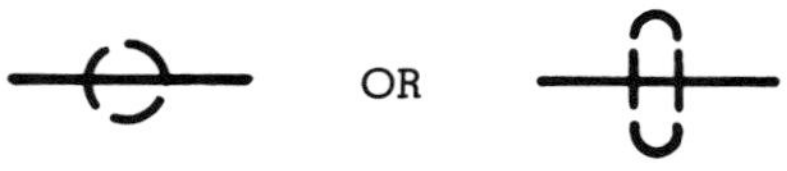

COMMONLY USED GRAPHIC SYMBOLS FOR ELECTRICAL AND ELECTRONIC DIAGRAMS	APPENDIX A

SHIELDED WIRE, TWO CONDUCTOR WITH
SHIELD GROUNDED

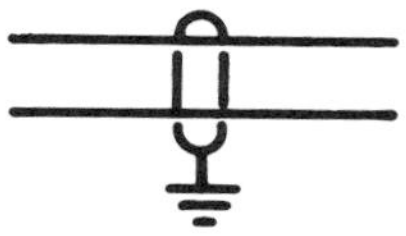

COAXIAL CABLE

GROUPING OF LEADS

Bend of line indicates direction of conductor joining group.

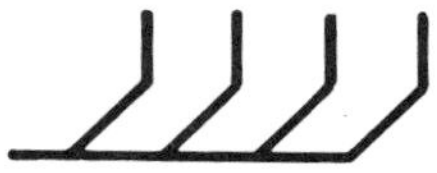

OR

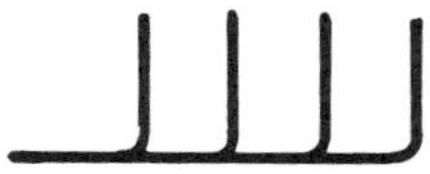

COMMON CONNECTIONS

All like designated points are assumed to be connected.

POSITIVE TERMINATION (ONE BATTERY)

NEGATIVE TERMINATION (ONE BATTERY)

POSITIVE TERMINATION (TWO BATTERIES)

NEGATIVE TERMINATION (TWO BATTERIES)

SINGLE GROUND TO EARTH

SEPARATE GROUNDS WITHIN A CIRCUIT OR SYSTEM

*P = Power Ground, S = Signal Ground

SINGLE CHASSIS OR FRAME CONNECTION

COMMONLY USED GRAPHIC SYMBOLS FOR ELECTRICAL AND ELECTRONIC DIAGRAMS	APPENDIX A

SEPARATE CHASSIS OR FRAME CONNECTIONS

    Note: This symbol indicates a conducting connection to the
          chassis or frame of a unit which may be at a substantial
          potential with respect to the earth or structure in which
          the chassis or frame is mounted.

MULTIPLE CONNECTIONS TO A COMMON VOLTAGE.
WHEN REQUIRED, BOTH POLARITIES OF A POWER
SUPPLY ARE SHOWN.

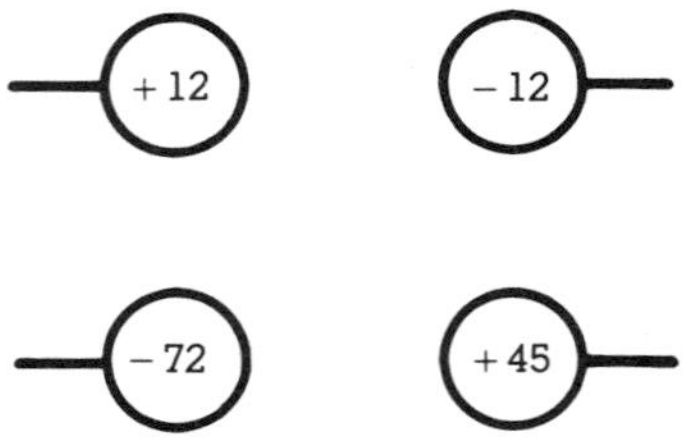

POLARITY MARKING
    POSITIVE

$+$

    NEGATIVE

$-$

LOGIC SYMBOLS

    "AND" GATE

    "OR" GATE (ALSO CALLED THE INCLUSIVE OR GATE)

    EXCLUSIVE "OR" GATE

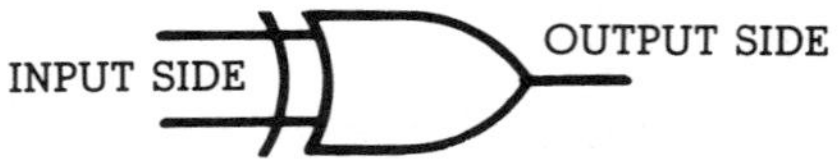

COMMONLY USED GRAPHIC SYMBOLS FOR ELECTRICAL AND ELECTRONIC DIAGRAMS	APPENDIX A

LOGIC NEGATION SYMBOL

The presence of a small circle (o) drawn at the point where a signal line joins a logic symbol indicates a logic negation or reversal of logic polarity of the signal appearing at an output or input.

FLIP-FLOP

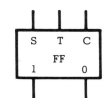

OR

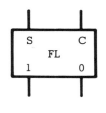

FLIP-FLOP LATCH

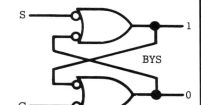

OR

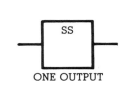

SINGLE-SHOT FUNCTIONS

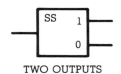

ONE OUTPUT

OR

TWO OUTPUTS

COMMONLY USED GRAPHIC SYMBOLS FOR ELECTRICAL AND ELECTRONIC DIAGRAMS	APPENDIX A

SCHMITT TRIGGER

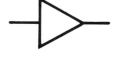

ONE OUTPUT

OR

TWO OUTPUTS

AMPLIFIER

Note: This symbol is pointed in the direction of signal transmission.

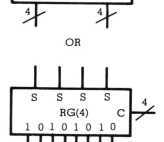

BINARY REGISTER

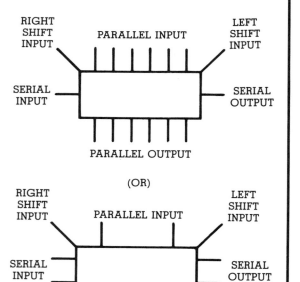

SHIFT REGISTER

Note: "Right Shift Input" is placed at a left corner of the symbol to indicate a shift from left to right. "Left Shift Input" is placed at a right corner of the symbol to indicate a shift from right to left.

COMMONLY USED GRAPHIC SYMBOLS FOR ELECTRICAL AND ELECTRONIC DIAGRAMS	APPENDIX A

MILITARY APPROVED ABBREVIATIONS FOR COMMONLY USED
WORDS AND WORD COMBINATIONS

GENERAL RULES

a. Abbreviations of word combinations shall not be separated for single use.

b. Single abbreviations may be combined when necessary.

c. Spaces between word combination abbreviations may be filled with a hyphen (-) for clarity.

d. The same abbreviations shall be used for all tenses, the possessive case, singular, and plural forms of a given word.

WORD	ABBREVIATION	WORD	ABBREVIATION
Absolute	ABS	April	APR
Accessory	ACCESS.	Arc Weld	ARC W
Accumulate	ACCUM	Armature	ARM.
Actual	ACT.	Assemble	ASSEM
Adapter	ADPT	Assembly	ASSY
Addendum	ADD.	Attention	ATTN
Addition	ADD.	Attenuation, Attenuator	ATTEN
Adjust	ADJ	Audio frequency	AF
After	AFT.	August	AUG
Alignment	ALIGN.	Authorize	AUTH
Allowance	ALLOW.	Automatic frequency control	AFC
Alloy	ALY	Automatic gain control	AGC
Alteration	ALT	Automatic volume control	AVC
Alternate	ALT	Auxiliary	AUX
Alternating current	AC	Auxiliary power unit	APU
Ambient	AMB	Average	AVG
American wire gage	AWG	Avoirdupois	AVDP
Amount	AMT	Azimuth	AZ
Ampere	AMP	Balance	BAL
Ampere (combination form)	A	Ball bearing	BB
Ampere hour	AMP HR	Bandpass	BP
Amplifier	AMPL	Bandwidth	B
Amplitude modulation	AM.	Base line	BL
And	&	Basic	BSC
Annunciator	ANN	Battery (electrical)	BAT.
Anodize	ANOD	Bearing	BRG.
Antenna	ANT.	Beat-frequency (adj)	B F
Application	APPL	Beat-frequency oscillator	BFO
Approved	APPD	Berylium	BE.
Approximate	APPROX	Between	BET.

WORD ABBREVIATIONS ON DRAWINGS	APPENDIX B

WORD	ABBREVIATION	WORD	ABBREVIATION
Between centers	BC	Change order	CO
Bill of Material	B/M	Chassis	CHAS
Binding	BIND.	Chromium	Cr
Black	BLK	Circuit	CKT
Blank	BLK	Circuit breaker	CKT BKR
Blower	BLO	Circular	CIR
Blue	BLU	Circular pitch	CP
Board	BD	Circumference	CIRC
Bolt circle	BC	Class	CL
Both faces	BF	Clearance	CL
Both sides	BS	Clockwise	CW
Bottom face	BF	Coaxial	COAX.
Bracket	BRKT	Coefficient	COEF
Brass	BRS	Cold-drawn steel	CDS
Brazing	BRZG	Cold-rolled	CR
Bridge	BRDG	Cold-rolled steel	CRS
Brinell hardness	BH	Collector	COLL
Brinell hardness number	BHN	Color code	CC
British thermal unit	BTU	Commercial	COML
Bronze	BRZ	Common	COM
Brown	BRN	Company	CO
Brown and Sharpe (gage)	B&S	Composition	COMP
Building	BLDG	Concentric	CONC
Burnish	BHN	Condition	COND
Bushing	BUSH.	Conductor	COND
Buzzer	BUZ	Connector	CONN
Bypass	BYP	Contact	CONT
Cadmium	CAD.	Continue	CONT
Calibrate	CAL	Continuous wave	CW
Camber	CAM.	Control	CONT
Capacitor	CAP.	Copper	COP
Capacity	CAP.	Corporation	CORP
Carbon	C	Corrosion	CORR
Carbon Steel	CS	Corrosion-resistant	CRE
Carton	CTN	Corrosion-resistant steel	CRES
Castellate	CTL	Cotangent	COT.
Casting	CSTG	Counterbore	CBORE
Cast Iron	CI	Counterclockwise	CCW
Cathode-ray tube	CRT	Counterdrill	CDRILL
Center	CTR	Countersink	CSK
Center line	CL or ¢	Countersink other side	CSKO
Center of gravity	CG	Cross section	XSECT
Center Tap	CT	Crystal	XTAL
Center to Center	C TO C	Cubic	CU
Centigrade	C	Cubic centimeter	CC
Centimeter	CM	Cubic feet	CU FT
Ceramic	CER	Cubic feet per minute	CFM
Chamfer	CHAM	Cubic feet per second	CFS
Change	CHG	Cubic inch	CU IN.
Change notice	CN	Cubic meter	CU M

WORD ABBREVIATIONS ON DRAWINGS	APPENDIX B

WORD	ABBREVIATION	WORD	ABBREVIATION
Cubic millimeter	CU MM	Engineer	ENGR
Current	CUR.	Engineering	ENGRG
Cycle	CY	Equipment	EQUIP.
Cycles per minute	CPM	Equivalent	EQUIV
Cycles per second	CPS	Estimate	EST
Datum	D	Et cetera	ETC
Decalcomania	DECAL	Example	EX
December	DEC	Exclusive	EXCL
Decibel	DB	Extension	EXT
Decimal	DEC	External	EXT
Deep-drawn	DD	Fahrenheit	F
Degree	DEG or °	Farad	F
Delay	DLY	Far side	FS
Delay line	DL	Fastener	FASTNR
Depth	D	February	FEB
Detail	DET	Federal	FED.
Deviation	DEV	Federal Specification	FS
Diagram	DIAG	Federal stock number	FSN
Diameter	DIA	Figure	FIG.
Diameter bolt circle	DBC	Filament	FIL
Diametral pitch	DP	Filament center tap	FCT
Dimension	DIM.	Fillister head	FIL H
Direct current	DC	Finish	FIN.
Direct-current volts	VDC	Finish all over	FAO
Direct-current working volts	VDCW	Fixed	FXD
Disconnect	DISC.	Flange	FLG
Division	DIV	Flat head	FH
Door	DR	Foot	FT or '
Double pole	DP	Foot-pound	FT LB
Double-pole double throw	DPDT	For example	EG
Double-pole single throw	DPST	Four-pole	4P
Double throw	DT	Frequency	FREQ
Down	DN	Frequency modulation	FM
Drawing	DWG	Front	FR
Drawing list	DL	Gage	GA
Drill	DR	Gallon	GAL
Drill rod	DR	Galvanize	GALV
Drive	DR	Half-hard	½ H
Drive fit	DF	Half-round	½ RD
Duplicate	DUP	Hard	H
Each	EA	Harden	HDN
Eccentric	ECC	Hardware	HDW
Effective	EFF	Head	HD
Electric	ELEC	Headless	HDLS
Electrolytic	ELECT.	Heater	HTR
Element	ELEM	Height	HGT
Eliminate	ELIM	Henry (electrical)	H
Elongation	ELONG	Hertz	HZ
Emergency	EMER	Hexagon	HEX.
End to end	E to E	Hexagonal head	HEX HD

WORD ABBREVIATIONS ON DRAWINGS	APPENDIX B

WORD	ABBREVIATION	WORD	ABBREVIATION
High frequency	HF	Machine screw	MS
High frequency oscillator	HFO	Maintenance	MAINT
Horizon, Horizontal	HORIZ	Major	MAJ
Horizontal center line	HCL	Manual	MAN
Horsepower	HP	Manufacture	MFR
Hot-rolled steel	HRS	Manufactured	MFD
Hour	HR	Manufacturing	MFG
Identification	IDENT	March	MAR.
Impeller	IMP.	Master oscillator	MO
Impulse	IMP.	Material	MATL
Inch	IN. or "	Maximum	MAX
Inches per second	IPS	Mechanical	MECH
Inch-pound	IN LB	Megacycle	MC
Include	INCL	Megacycle per second	MCS
Inclusive	INCL	Megohm	MEGO
Incorporated	INC	Meter	M
Indicate	IND	Microfarad	UF or uF
Indicator	IND	Microhenry	UH or uH
Information	INFO	Microhm	UOHM or
Inside diameter	ID		uOHM
Inside radius	IR	Microinch	UIN or uIN
Insulation, Insulator	INS	Micro micro (10^{-12})	UU or uu
Intermediate frequency	IF	Micromicrohenry	UUH or uuH
Internal	INT	Micromicrofarad	UUF or uuF
Irregular	IRREG	Micromicron	UU or uu
Issue	ISS	Micron (.001 millimeter)	U or u
Jack	J	Milliampere	MA
January	JAN	Millihenry	MH
July	JUL	Millimeter	MM
Junction	JCT	Millivolt	MV
Junction box	JB	Milliwatt	MW
June	JUN	Minimum	MIN
Keyway	KWY	Minor	MIN
Kilocycle	KC	Miscellaneous	MISC
Kilocycles per second	KC/S	Model (for general use)	MOD
Left	L	Modification	MOD
Left hand	LH	Modify	MOD
Length	LG	Modulator	MOD
Length over-all	LOA	Mount	MT
Light	LT	Mounting	MTG
Linear	LIN	National coarse (thread)	NC
Liquid	LIQ	National extra fine (thread)	NEF
Load limiting resistor	LLR	National fine (thread)	NF
Local oscillator	LO	National special (thread)	NS
Lock washer	LK WASH.	Nickel	NI
Long	LG	Nominal	NOM
Loudspeaker	LS	Normal	NORM.
Low frequency	LF	Normally closed	NC
Low-frequency oscillator	LFO	Normally open	NO.
Low pass	LP	Not applicable	NA

WORD ABBREVIATIONS ON DRAWINGS	APPENDIX B

WORD	ABBREVIATION	WORD	ABBREVIATION
Not to scale	NTS	Pound-foot	LB FT
November	NOV	Pounds per square inch	PSI
Number	NO.	Power	PWR
Nylon	N	Power amplifier	PA
Obsolete	OBS	Power supply	PWR SUP
October	OCT	Preamplifier	PREAMP
Ohm (for use only on diagrams)	Ω	Preferred	PFD
On center	OC	Preliminary	PRELIM
Opposite	OPP	Prepare	PREP
Optional	OPT	Primary	PRIM.
Orange	ORN	Project	PROJ
Origin, Original	ORIG	Quantity	QTY
Oscillator	OSC	Quarter-hard	¼ H
Ounce	OZ	Quartz	QTZ
Ounce-inch	OZ IN	Radio frequency	RF
Output	OUPT	Radius	R or RAD.
Outside diameter	OD	Rate	RT
Outside radius	OR.	Reactor	REAC
Oval head	OV HD	Received	RECD
Page	P	Receiver	RCVR
Pair	PR	Receptable	RECP
Panel	PNL	Rectifier	RECT
Paragraph	PARA	Reference	REF
Parallel	PAR.	Reference line	REF L
Part	PT	Relay	REL
Part number	PN	Release	REL
Passivate	PASS.	Relief	REL
Per	/	Remove	REM
Per centum	PCT	Required	REQD
Perpendicular	PERP	Resistor	RES.
Phase	PH	Revision	REV
Phenolic	PHEN	Rheostat	RHEO
Phillips head	PHL H	Right	R
Phosphor bronze	PH BRZ	Right hand	RH
Piece	PC	Rivet	RIV
Pitch	P	Rockwell hardness	RH
Pitch circle	PC	Rotary	ROT.
Pitch diameter	PD	Round	RD
Plastic	PLSTC	Roundhead	RH
Plate	PL	Schedule	SCH
Plate (electron tube)	P	Schematic	SCHEM
Plug	PL	Screw	SCR
Plus or minus	±	Seamless	SMLS
Point	PT	Seamless steel tubing	SSTU
Point of intersection	PI	Section	SECT.
Point of tangency	PT	Selector	SEL
Pole	P	September	SEP
Position	POS	Serial	SER
Positive	POS	Servo	SVO
Pound	LB	Set screw	SS

WORD ABBREVIATIONS ON DRAWINGS	APPENDIX B

WORD	ABBREVIATION	WORD	ABBREVIATION
Sheet	SH	Three-pole	3P
Shield	SHLD	Through	THRU
Shoulder	SHLD	Time delay	TD
Silver	SIL	Toggle	TGL
Similar	SIM	Tolerance	TOL
Single pole	SP	Torque	TOR
Single pole, double throw	SPDT	Transformer	XMFR
Single pole, single throw	SPST	Transistor	TSTR
Single throw	ST	Transmitter	XMTR
Slotted	SLOT.	Transmitter receiver	TR
Small	SM	Triple pole	3P
Socket	SOC	Triple throw	3T
Socket head	SCH	True position	TP
Solenoid	SOL.	Tubing	TUB.
Spacer	SPR	Twisted	TW
Speaker	SPKR	Typical	TYP
Special	SPL	Unfinished	UNFIN
Specification	SPEC	Unified coarse thread	UNC
Spectrum analyzer	SA	Unified extra fine thread	UNEF
Spherical	SPHER	Unified fine thread	UNF
Split ring	SR	Unified special thread	UNS
Spot face	SF	United States Air Force	USAF
Spot weld	SW	Upper	UP.
Spring	SPG	Used with	U/W
Square	SQ	Variable	VAR
Square foot	SQ FT	Variable frequency oscillator	VFO
Square inch	SQ IN.	Vertical	VERT.
Stainless steel	SST	Vertical center line	VCL
Standard	STD	Very-high frequency	VHF
Steel	STL	Very-low frequency	VLF
Stranded	STRD	Video	VID
Surface	SURF.	Viscosity	VIS
Switch	SW	Volt	V
Symbol	SYM	Washer	WASH.
Symmetrical	SYM	Watt	W
Tangent	TAN.	Weight	WT
Tapping	TAP.	White	WHT
Tempered	TEMP	Wire-wound	WW
Terminal	TERM.	With (abbreviate only in	
That is	i.e.	conjunction with other	
Thermal	THRM	abbreviations)	W/
Thermistor	TMTR	Without	W/O
Thick	THK	Yellow	YEL
Thread	THD	Zinc	Zn
Threads per inch	TPI		

CLASS DESIGNATION LETTERS

FROM MIL STD 16B

Alarm	DS	Junction, hybrid	HY
Amplifier	A	Key, switch	S
Amplifier, rotating	G	Lamp, pilot or illuminating	DS
Annunciator	DS	Lamp, signal	DS
Antenna, aerial	E	Line, delay	DL
Arrestor, lightning	E	Loop antenna	E
Assembly	A	Magnet	E
Attenuator	AT	Meter	M
Audible signaling device	DS	Microphone	MK
Autotransformer	T	Mode transducer	MT
Battery	BT	Modulator	A
Bell	DS	Motor	B
Blower, fan, motor	B	Motor-generator	MG
Board, terminal	TB	Mounting (not in electric circuit and	
Breaker, circuit	CB	not in a socket)	MP
Buzzer	DS	Nameplate	N
Cable	W	Oscillator (excluding elect. tube	
Capacitor	C	used in oscillator)	Y
Cell, aluminum or electrolytic	E	Oscilloscope	M
Cell, light-sensitive, photoemissive	V	Pad	AT
Choke	L	Part, miscellaneous	E
Circuit breaker	CB	Path, guided, transmission	W
Coil, hybrid	HY	Phototube	V
Coil, induction, relay tuning, operating	L	Pickup, erasing, recording, or	
Coil, repeating	T	reproducing head	PU
Computer	A	Plug (see connector)	
Connector, receptacle, affixed to wall,		Potentiometer	R
chassis, panel	J	Power supply	A
Connector, receptacle, affixed to		Receiver, telephone	HT
end of cable, wire	P	Receptacle (fixed connector)	J
Contact, electrical	E	Rectifier, crystal or metallic	CR
Contactor, electrically operated	K	Regulator, voltage (except electron tube)	VR
Contactor, mechanically or		Relay, electrically operated contactor	
thermally operated	S	or switch	K
Coupler, directional	DS	Repeater (telephone usage)	RP
Crystal detector	CR	Resistor	R
Crystal diode	CR	Rheostat	R
Crystal, piezoelectric	Y	Selenium cell	CR
Cutout, fuse	F	Shunt	R
Cutout, thermal	S	Solenoid	L
Detector, crystal	CR	Speaker	LS
Device, indicating	DS	Speed regulator	S
Dipole antenna	E	Strip, terminal	TP
Disconnecting device	S	Subassembly	A
Electron tube	V	Switch, mechanically or	
Exciter	G	thermally operated	S
Fan	B	Terminal board or strip	TB
Filter	FL	Test point	TP
Fuse	F	Thermistor	RT
Generator	G	Thermocouple	TC
Handset	HS	Thermostat	S
Head, erasing, recording, reproducing	PU	Timer	M
Heater	HR	Transducer	MT
Horn, howler	LS	Transformer	T
Indicator	DS	Transistor	Q
Inductor	L	Transmission path	W
Instrument	M	Tube, electron	V
Insulator	E	Varistor, asymmetrical	CR
Integrated circuit package	U	Varistor, symmetrical	RV
Interlock, mechanical	MP	Voltage regulator (except an electron	
Interlock, safety, electrical	S	tube)	VR
Jack (see connector, receptacle, electrical)		Waveguide	W
Junction, coaxial or waveguide		Winding	L
(tee or wye)	CP	Wire	W

CLASS DESIGNATION LETTERS	APPENDIX C

RESISTOR AND CAPACITOR STANDARD COLOR CODE

STANDARD COLOR CODE FOR RESISTORS AND CAPACITORS

The standard color code provides the necessary information required to properly identify color coded resistors and capacitors. Refer to the color code for numerical value and the number of zeroes (or multiplier) assigned to the colors used. A fourth color band on resistors determines the tolerance rating. Absence of the fourth band indicates a 20% tolerance rating. (REF MIL-STD-221)

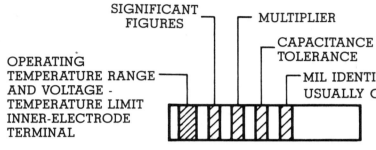

AXIAL LEAD RESISTOR

TOLERANCE

MULTIPLIER

1ST AND 2ND SIGNIFICANT FIGURES

WIRE-WOUND RESISTORS HAVE FIRST DIGIT BAND DOUBLE WIDTH

CAPACITORS

SIGNIFICANT FIGURES

MULTIPLIER

CAPACITANCE TOLERANCE

MIL IDENTIFIER USUALLY ORANGE

OPERATING TEMPERATURE RANGE AND VOLTAGE - TEMPERATURE LIMIT INNER-ELECTRODE TERMINAL

NOTE: THE COLOR RINGS MAY BE DOTS.

COLOR CODE

COLOR	FIRST FIGURE	SECOND FIGURE	MULTIPLIER	TOLERANCE	
				%	LTR
BLACK	0	0	1		
BROWN	1	1	10		
RED	2	2	100	± 2%	G
ORANGE	3	3	1,000		
YELLOW	4	4	10,000		
GREEN	5	5	100,000		
BLUE	6	6	1,000,000		
PURPLE (VIOLET)	7	7			
GRAY	8	8			
WHITE	9	9			
SILVER			0.01	± 10%	K
GOLD			0.1	± 5%	J
				± 20%	M
				± 1%	F

CONDUCTOR COLOR ABBREVIATIONS

COLOR	PREFERRED ABBREVIATION	COLOR	PREFERRED ABBREVIATION
BLACK	BK	BLUE	BL
BROWN	BR	VIOLET	V
RED	R	(PURPLE)	(PR)
ORANGE	O	GRAY	GY
YELLOW	Y	(SLATE)	(S)
GREEN	G	WHITE	W

COLOR CODING AND COLOR ABBREVIATIONS	APPENDIX D

WIRE AND SHEET GAGE

NUMBER GAGE	*AMERICAN WIRE GAGE OR BROWN & SHARPE GAGE	**UNITED STATES STANDARD GAGE	MACHINE AND WOOD SCREW GAGE
6/0	.5800	—	—
5/0	.5165	—	—
4/0	.4600	.4063	—
3/0	.4096	.3750	—
2/0	.3648	.3438	—
0	.3249	.3125	.060
1	.2893	.2813	.073
2	.2576	.2656	.086
3	.2294	.2500	.099
4	.2043	.2344	.112
5	.1819	.2188	.125
6	.1620	.2031	.138
7	.1443	.1875	.151
8	.1285	.1719	.164
9	.1144	.1563	.177
10	.1019	.1406	.190
11	.0907	.1250	.203
12	.0808	.1094	.216
13	.0720	.0938	—
14	.0641	.0781	.242
15	.0571	.0703	—
16	.0508	.0625	.268
17	.0453	.0563	—
18	.0403	.0500	.294
19	.0359	.0438	—
20	.0320	.0375	.320
21	.0285	.0344	—
22	.0253	.0313	—
23	.0226	.0281	—
24	.0201	.0250	.372
25	.0179	.0219	—
26	.0159	.0188	—
27	.0142	.0172	—
28	.0126	.0156	—
29	.0113	.0141	—
30	.0100	.0125	.450
31	.0089	.0109	—
32	.0080	.0102	—
33	.0071	.0094	—
34	.0063	.0086	—
35	.0056	.0078	—
36	.0050	.0070	—

* American or Brown & Sharpe Gage: copper wire, brass, copper alloys and nickel silver wire and sheet, also aluminum sheet, rod, and wire.

**United States Standard Gage: steel and Monel metal sheets.
The use of decimals of an inch for dimensions specifying sheet and wire is recommended.

DECIMAL EQUIVALENTS - WIRE AND SHEET GAGE	APPENDIX E

DECIMAL EQUIVALENTS

FRACTION			DECIMAL		FRACTION			DECIMAL	
			INCH	METRIC (mm)				INCH	METRIC (mm)
		1/64	.015625	0.3969			33/64	.515625	13.0969
	1/32		.03125	0.7938		17/32		.53125	13.4938
		3/64	.046875	1.1906			35/64	.546875	13.8906
1/16			.0625	1.5875	9/16			.5625	14.2875
		5/64	.078125	1.9844			37/64	.578125	14.6844
	3/32		.09375	2.3813		19/32		.59375	15.0813
		7/64	.109375	2.7781			39/64	.609375	15.4781
1/8			.1250	3.1750	5/8			.6250	15.8750
		9/64	.140625	3.5719			41/64	.640625	16.2719
	5/32		.15625	3.9688		21/32		.65625	16.6688
		11/64	.171875	4.3656			43/64	.671875	17.0656
3/16			.1875	4.7625	11/16			.6875	17.4625
		13/64	.203125	5.1594			45/64	.703125	17.8594
	7/32		.21875	5.5563		23/32		.71875	18.2563
		15/64	.234375	5.9531			47/64	.734375	18.6531
1/4			.250	6.3500	3/4			.750	19.0500
		17/64	.265625	6.7469			49/64	.765625	19.4469
	9/32		.28125	7.1438		25/32		.78125	19.8438
		19/64	.296875	7.5406			51/64	.796875	20.2406
5/16			.3125	7.9375	13/16			.8125	20.6375
		21/64	.328125	8.3384			53/64	.828125	21.0344
	11/32		.34375	8.7313		27/32		.84375	21.4313
		23/64	.359375	9.1281			55/64	.859375	21.8281
3/8			.3750	9.5250	7/8			.8750	22.2250
		25/64	.390625	9.9219			57/64	.890625	22.6219
	13/32		.40625	10.3188		29/32		.90625	23.0188
		27/64	.421875	10.7156			59/64	.921875	23.4156
7/16			.4375	11.1125	15/16			.9375	23.8125
		29/64	.453125	11.5094			61/64	.953125	24.2094
	15/32		.46875	11.9063		31/32		.96875	24.6063
		31/64	.484375	12.3031			63/64	.984375	25.0031
1/2			.500	12.7000	1			1.000	25.4000

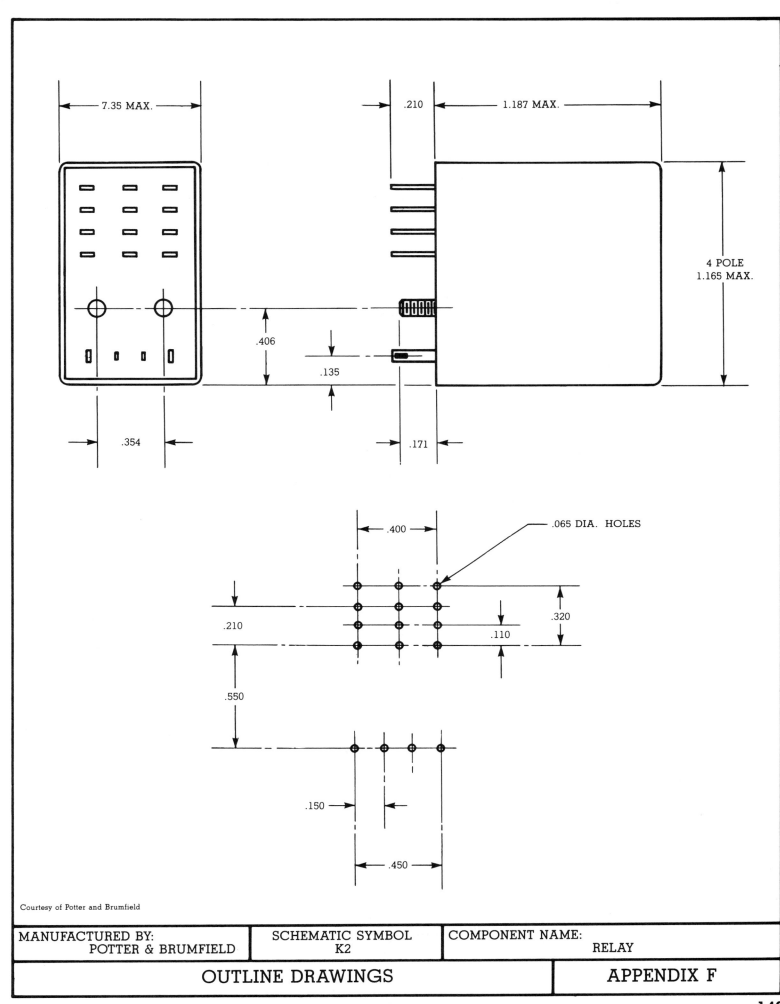

7.35 MAX.

.210

1.187 MAX.

4 POLE
1.165 MAX.

.406

.135

.354

.171

.400

.065 DIA. HOLES

.210

.320

.110

.550

.150

.450

Courtesy of Potter and Brumfield

MANUFACTURED BY: POTTER & BRUMFIELD	SCHEMATIC SYMBOL K2	COMPONENT NAME: RELAY
OUTLINE DRAWINGS		APPENDIX F

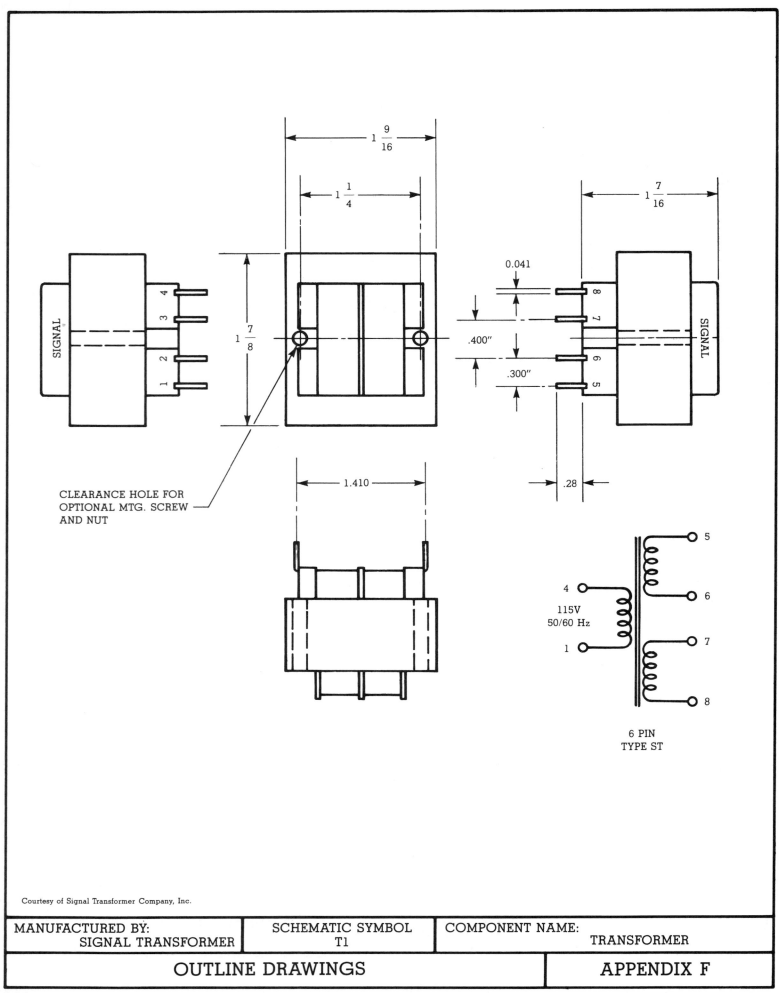

1 9/16

1 1/4

1 7/16

SIGNAL

4
3
2
1

0.041

8
7
6
5

SIGNAL

1 7/8

.400"

.300"

.28

CLEARANCE HOLE FOR
OPTIONAL MTG. SCREW
AND NUT

1.410

4
115V
50/60 Hz
1

5

6

7

8

6 PIN
TYPE ST

MANUFACTURED BY: SIGNAL TRANSFORMER	SCHEMATIC SYMBOL T1	COMPONENT NAME: TRANSFORMER
OUTLINE DRAWINGS		APPENDIX F

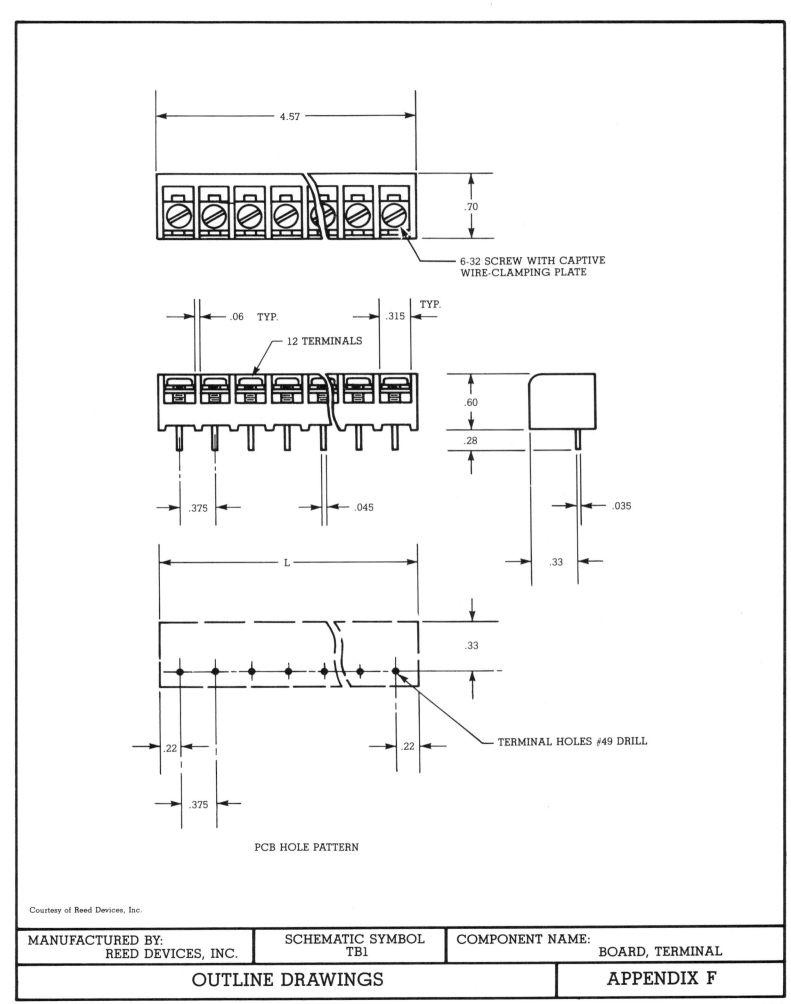

4.57

.70

6-32 SCREW WITH CAPTIVE
WIRE-CLAMPING PLATE

.06 TYP.

.315 TYP.

12 TERMINALS

.60

.28

.375

.045

.035

.33

L

.33

.22

.22

TERMINAL HOLES #49 DRILL

.375

PCB HOLE PATTERN

MANUFACTURED BY: REED DEVICES, INC.	SCHEMATIC SYMBOL TB1	COMPONENT NAME: BOARD, TERMINAL
OUTLINE DRAWINGS		APPENDIX F

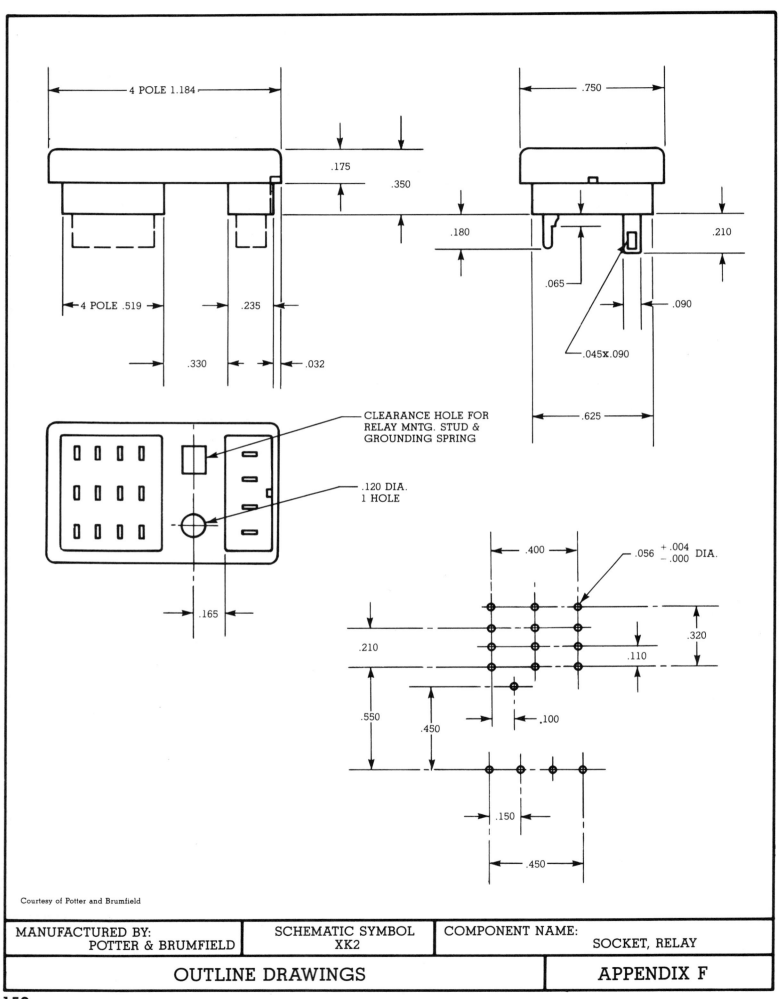

4 POLE 1.184

.175
.350
.180

4 POLE .519

.235

.330
.032

.750

.210

.065

.090

.045x.090

.625

CLEARANCE HOLE FOR
RELAY MNTG. STUD &
GROUNDING SPRING

.120 DIA.
1 HOLE

.165

.400

$.056 \begin{smallmatrix} +.004 \\ -.000 \end{smallmatrix}$ DIA.

.320

.110

.210

.550

.450

.100

.150

.450

MANUFACTURED BY: POTTER & BRUMFIELD	SCHEMATIC SYMBOL XK2	COMPONENT NAME: SOCKET, RELAY
OUTLINE DRAWINGS		APPENDIX F

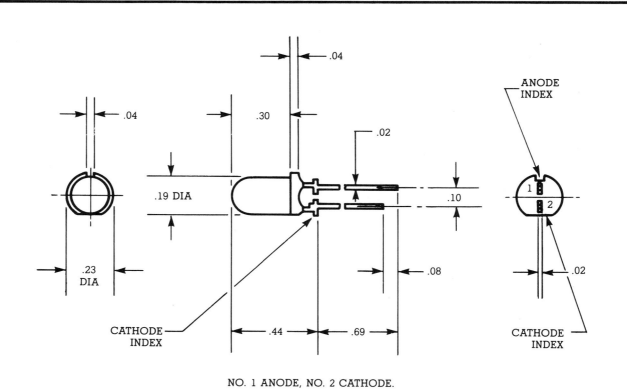

.04

.30

.04

.02

.19 DIA

.10

.23
DIA

.08

ANODE
INDEX

1

2

CATHODE
INDEX

.02

CATHODE
INDEX

.44

.69

NO. 1 ANODE, NO. 2 CATHODE.

MANUFACTURED BY: GENERAL INSTRUMENT	SCHEMATIC SYMBOL LED 1	COMPONENT NAME: DIODE, LIGHT EMITTING, RED

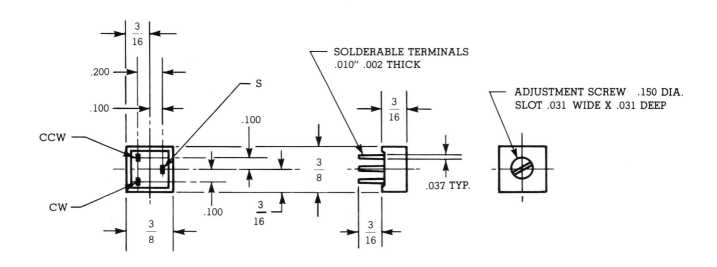

$\frac{3}{16}$

.200

.100

S

SOLDERABLE TERMINALS
.010" .002 THICK

$\frac{3}{16}$

ADJUSTMENT SCREW .150 DIA.
SLOT .031 WIDE X .031 DEEP

.100

CCW

$\frac{3}{8}$

CW

$\frac{3}{8}$

.100

$\frac{3}{16}$

.037 TYP.

$\frac{3}{16}$

MANUFACTURED BY: BECKMAN INSTRUMENTS	SCHEMATIC SYMBOL R2	COMPONENT NAME: RESISTOR, VARIABLE - 25 K OHM, 1/2W
OUTLINE DRAWINGS		APPENDIX F

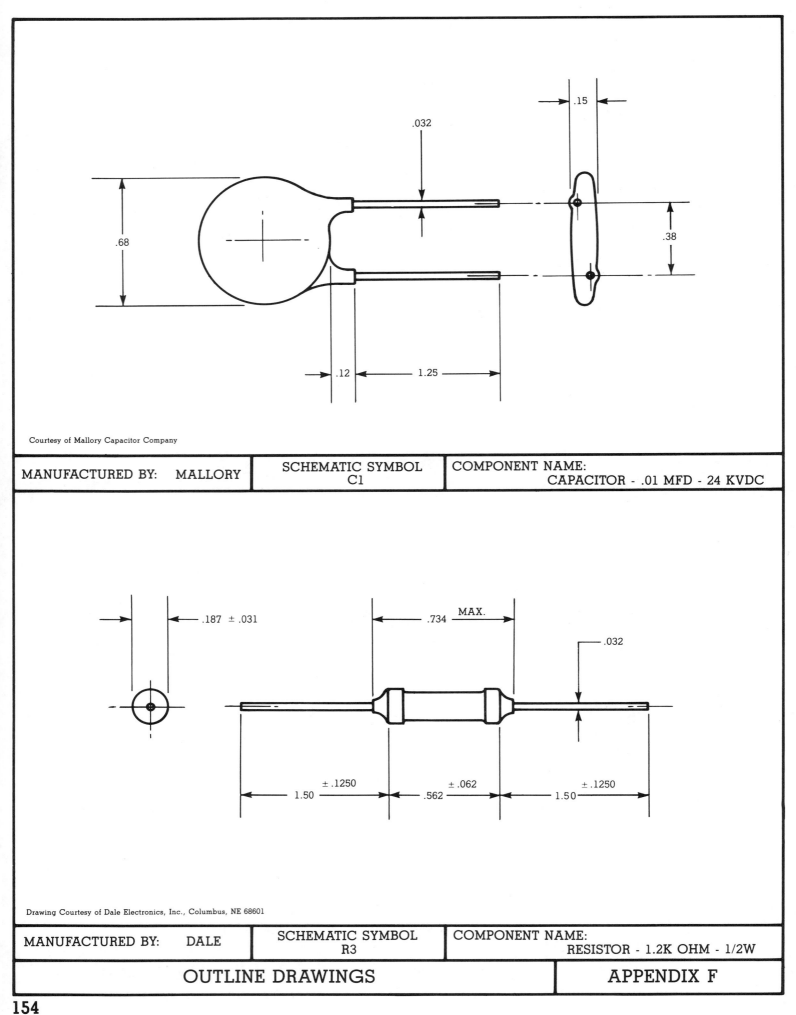

.032

.15

.68

.38

.032

.12

1.25

Courtesy of Mallory Capacitor Company

MANUFACTURED BY: MALLORY	SCHEMATIC SYMBOL C1	COMPONENT NAME: CAPACITOR - .01 MFD - 24 KVDC

.187 ± .031

.734   MAX.

.032

± .1250
1.50

± .062
.562

± .1250
1.50

Drawing Courtesy of Dale Electronics, Inc., Columbus, NE 68601

MANUFACTURED BY: DALE	SCHEMATIC SYMBOL R3	COMPONENT NAME: RESISTOR - 1.2K OHM - 1/2W

OUTLINE DRAWINGS	APPENDIX F

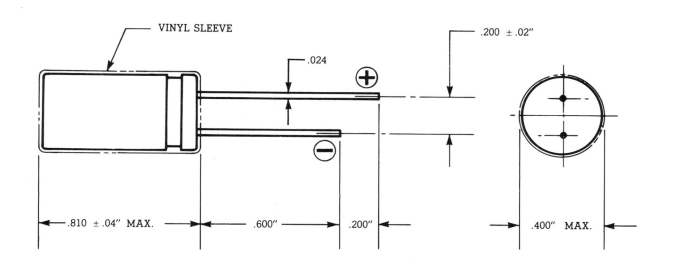

VINYL SLEEVE

.024

.200 ± .02"

⊕

⊖

.810 ± .04" MAX. ← .600" → .200" → .400" MAX.

Courtesy of Sprague Electronics

MANUFACTURED BY: SPRAGUE	SCHEMATIC SYMBOL C2, C3	COMPONENT NAME: CAPACITOR, ELECTROLYTIC, 100 MFD-50VDC

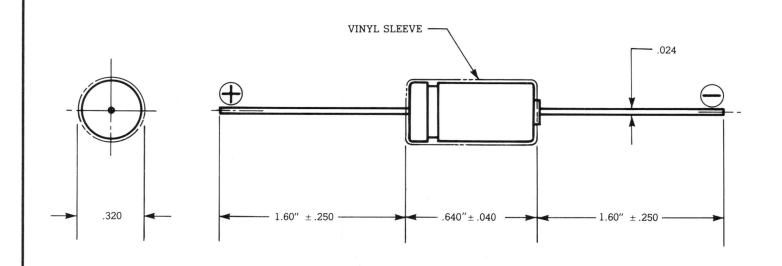

VINYL SLEEVE

.024

⊕

⊖

.320 ← 1.60" ± .250 → .640" ± .040 → 1.60" ± .250

Courtesy of Sprague Electronics

MANUFACTURED BY: SPRAGUE	SCHEMATIC SYMBOL C4	COMPONENT NAME: CAPACITOR, ELECTROLYTIC-10MFD-50VDC

OUTLINE DRAWINGS	APPENDIX F

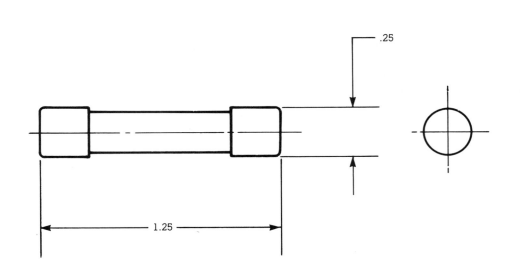

.25

1.25

MANUFACTURED BY: BUSSMAN	SCHEMATIC SYMBOL F1	COMPONENT NAME: FUSE - 1/2 AMP - 250V

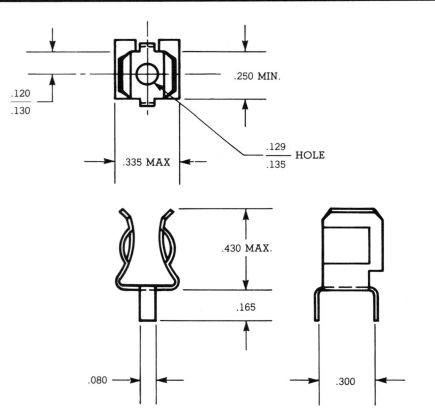

.250 MIN.

.120
.130

.129
.135 HOLE

.335 MAX

.430 MAX.

.165

.080

.300

MANUFACTURED BY: BUSSMAN	SCHEMATIC SYMBOL XF1	COMPONENT NAME: CLIP, FUSE
OUTLINE DRAWINGS		APPENDIX F

156

PERFORMANCE SPECIFICATIONS

1.0 COIL
   a. Resistance 2500 ohms
   b. Nominal Operate 100 V.D.C.
   c. Must Operate 8.55 V.D.C.
   d. Must Not Operate 3.75 V.D.C.
   e. Must Release .94 V.D.C.

2.0 CONTACT RATING
   a. 0.500 Amp. Maximum
   b. 50 D.C. Volts Maximum
   c. .200 ohms Contact Resistance
   d. Rhodium Contact Material

SCHEMATIC

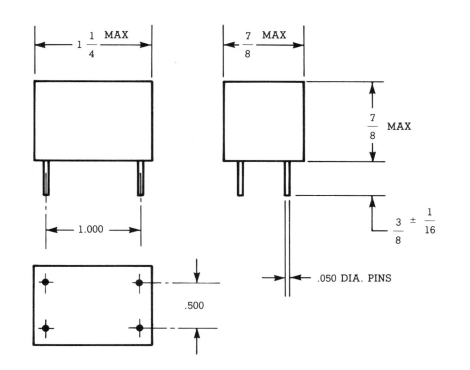

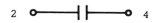

Courtesy of Wabash Relay & Electronics Company

MANUFACTURED BY: WABASH RELAY & ELECTRONICS	SCHEMATIC SYMBOL K1	COMPONENT NAME: RELAY, REED

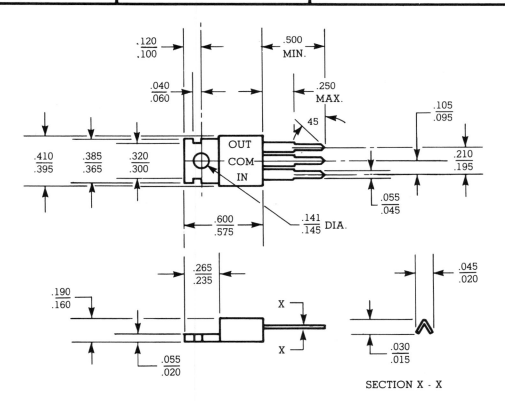

SECTION X - X

ᶜFairchild Camera and Instrument Corporation

Reprinted with permission of Fairchild Camera and Instrument Corporation

MANUFACTURED BY: FAIRCHILD	SCHEMATIC SYMBOL REG. 1	COMPONENT NAME: REGULARTOR, VOLTAGE - 24 VDC

OUTLINE DRAWINGS	APPENDIX F

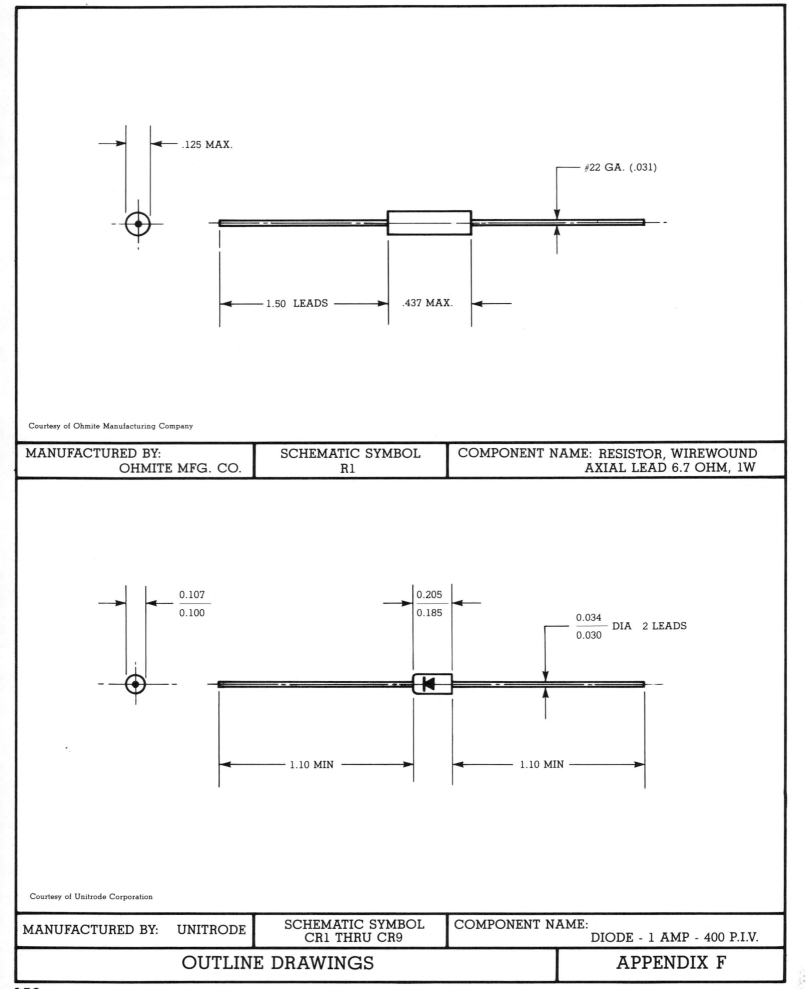

.125 MAX.

#22 GA. (.031)

1.50 LEADS .437 MAX.

MANUFACTURED BY: OHMITE MFG. CO.	SCHEMATIC SYMBOL R1	COMPONENT NAME: RESISTOR, WIREWOUND AXIAL LEAD 6.7 OHM, 1W

0.107 / 0.100

0.205 / 0.185

0.034 / 0.030 DIA   2 LEADS

1.10 MIN 1.10 MIN

MANUFACTURED BY: UNITRODE	SCHEMATIC SYMBOL CR1 THRU CR9	COMPONENT NAME: DIODE - 1 AMP - 400 P.I.V.
OUTLINE DRAWINGS		APPENDIX F

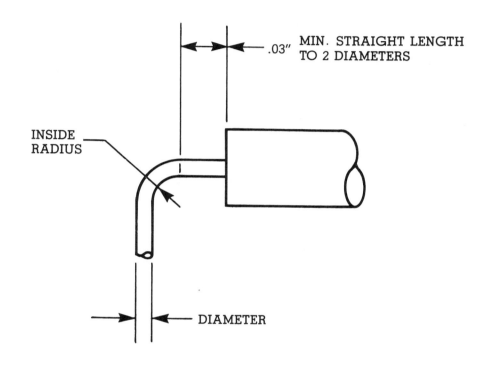

.03″ MIN. STRAIGHT LENGTH TO 2 DIAMETERS

INSIDE RADIUS

DIAMETER

LEAD DIAMETER	MINIMUM INSIDE RADIUS
UP TO .027 INCHES	ONE DIAMETER UP TO .060 INCHES
FROM .028 TO .047 INCHES	1.5 DIAMETERS UP TO .060 INCHES
.048 INCHES AND GREATER	2 DIAMETERS
BENDS - COMPONENT LEADS	APPENDIX G

PREFERRED BEND RADII FOR STRAIGHT BENDS IN SHEET METALS

SHEET THICKNESS	ALUMINUM						
	2024 T3	5052 0	5052 H32	5052 H34	6061 0	6061 T4	6061 T6
.020	.06	.03	.03	.03	.03	.06	.06
.025	.06	.03	.03	.03	.03	.06	.06
.032	.09	.03	.06	.06	.03	.09	.09
.040	.12	.06	.06	.06	.06	.12	.12
.050	.16	.06	.09	.09	.06	.16	.16
.063	.19	.06	.09	.09	.06	.19	.19
.080	.25	.09	.12	.12	.09	.25	.25
.090	.31	.09	.12	.12	.09	.31	.31
.100	.38	.12	.19	.19	.12	.38	.38
.125	.50	.12	.19	.19	.12	.50	.50
.160	.75	.16	.25	.25	.16	.62	.62
.190	1.00	.19	.37	.37	.19	.84	.87

SHEET THICKNESS	STEEL		
	CORROSION RESISTANT		PLAIN CARBON
	TYPES 301 302 304 (Annealed)	TYPES 301 302 304 (1/4 H)	
.020			.06
.025			.06
.032			.06
.040			.09
.050			.09
.063			.09
.080			.12
.090			.12
.100			.16
.125			.19
.160			.25
.190			.31
.020-.040	.03	.06	
.045-.070	.06	.12	
.075-.105	.09	.19	
.110-.135	.12	.25	

BENDS - RADII IN SHEET METALS	APPENDIX G

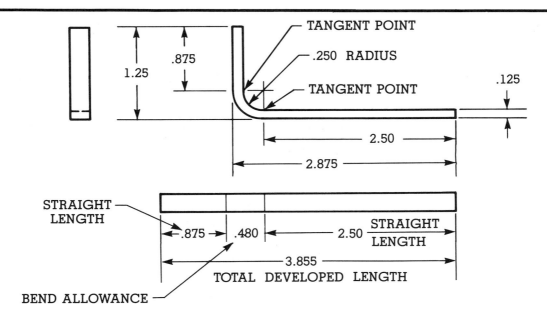

To determine the developed length of an object with a 90 degree bend.

1. Locate the tangent points at the inside of the .250 inch radius. The tangent point is where the straight length meets the radius.
2. Find the material thickness on the drawing: .125 inch.
3. Refer to the chart below where the horizontal row at the top indicates the inside radius and the left hand column indicates the material thickness. The chart shows the .250 radius and the .125 thickness meet at .480.

This dimension is called the bend allowance and is the length that must be added to the straight lengths to determine the total developed length.

BEND ALLOWANCE FOR 90° BENDS (INCH)

Radii Thickness	.031	.063	.094	.125	.156	.188	.219	.250	.281	.313	.344	.375	.438	.500
.013	.058	.108	.157	.205	.254	.304	.353	.402	.450	.501	.549	.598	.697	.794
.016	.060	.110	.159	.208	.256	.307	.355	.404	.453	.503	.552	.600	.699	.796
.020	.062	.113	.161	.210	.259	.309	.358	.406	.455	.505	.554	.603	.702	.799
.022	.064	.114	.163	.212	.260	.311	.359	.408	.457	.507	.556	.604	.703	.801
.025	.066	.116	.165	.214	.263	.313	.362	.410	.459	.509	.558	.607	.705	.803
.028	.068	.119	.167	.216	.265	.315	.364	.412	.461	.511	.560	.609	.708	.805
.032	.071	.121	.170	.218	.267	.317	.366	.415	.463	.514	.562	.611	.710	.807
.038	.075	.126	.174	.223	.272	.322	.371	.419	.468	.518	.567	.616	.715	.812
.040	.077	.127	.176	.224	.273	.323	.372	.421	.469	.520	.568	.617	.716	.813
.050		.134	.183	.232	.280	.331	.379	.428	.477	.527	.576	.624	.723	.821
.064		.144	.192	.241	.290	.340	.389	.437	.486	.536	.585	.634	.732	.830
.072			.198	.247	.296	.346	.394	.443	.492	.542	.591	.639	.738	.836
.078			.202	.251	.300	.350	.399	.447	.496	.546	.595	.644	.743	.840
.081			.204	.253	.302	.352	.401	.449	.498	.548	.598	.646	.745	.842
.091			.212	.260	.309	.359	.408	.456	.505	.555	.604	.653	.752	.849
.094			.214	.262	.311	.361	.410	.459	.507	.558	.606	.655	.754	.851
.102				.268	.317	.367	.416	.464	.513	.563	.612	.661	.760	.857
.109				.273	.321	.372	.420	.469	.518	.568	.617	.665	.764	.862
►.125				.284	.333	.383	.432	►.480	.529	.579	.628	.677	.776	.873
.156					.355	.405	.453	.502	.551	.601	.650	.698	.797	.895
.188						.427	.476	.525	.573	.624	.672	.721	.820	.917
.203								.535	.584	.634	.683	.731	.830	.928
.218								.546	.594	.645	.693	.742	.841	.938
.234								.557	.606	.656	.705	.753	.852	.950
.250								.568	.617	.667	.716	.764	.863	.961

BENDS - 90 DEGREE DEVELOPED LENGTH	APPENDIX G

DRILL SIZE CHART

DRILL SIZE	DECIMAL DIA.	DRILL SIZE	DECIMAL DIA.	DRILL SIZE	DECIMAL DIA.	DRILL SIZE	DECIMAL DIA.	DRILL SIZE	DECIMAL DIA
80	.0135	50	.0700	22	.1570	17/64	.2656	1/2	.5000
79	.0145	49	.0730	21	.1590	H	.2660	33/64	.5156
1/64	.0156	48	.0760	20	.1610	I	.2720	17/32	.5312
78	.0160	5/64	.0781	19	.1660	J	.2770	35/64	.5469
77	.0180	47	.0785	18	.1695	K	.2811	9/16	.5625
76	.0200	46	.0810	11/64	.1719	9/32	.2812	37/64	.5781
75	.0210	45	.0820	17	.1730	L	.2900	19/32	.5937
74	.0225	44	.0860	16	.1770	M	.2950	39/64	.6094
73	.0240	43	.0890	15	.1800	19/64	.2968	5/8	.6250
72	.0250	42	.0935	14	.1820	N	.3020	41/64	.6406
71	.0260	3/32	.0937	13	.1850	5/16	.3125	21/32	.6562
70	.0280	41	.0960	3/16	.1875	O	.3160	43/64	.6719
69	.0292	40	.0980	12	.1890	P	.3230	11/16	.6875
68	.0310	39	.0995	11	.1910	21/64	.3281	45/64	.7031
1/32	.0313	38	.1015	10	.1935	Q	.3320	23/32	.7187
67	.0320	37	.1040	9	.1960	R	.3390	47/64	.7344
66	.0330	36	.1065	8	.1990	11/32	.3437	3/4	.7500
65	.0350	7/64	.1093	7	.2010	S	.3480	49/64	.7656
64	.0360	35	.1100	13/64	.2031	T	.3580	25/32	.7812
63	.0370	34	.1110	6	.2040	23/64	.3594	51/64	.7969
62	.0380	33	.1130	5	.2055	U	.3680	13/16	.8125
61	.0390	32	.1160	4	.2090	3/8	.3750	53/64	.8281
60	.0400	31	.1200	3	.2130	V	.3770	27/32	.8437
59	.0410	1/8	.1250	7/32	.2187	W	.3860	55/64	.8594
58	.0420	30	.1285	2	.2210	25/64	.3906	7/8	.8750
57	.0430	29	.1360	1	.2280	X	.3970	57/64	.8906
56	.0465	28	.1405	A	.2340	Y	.4040	29/32	.9062
3/64	.0469	9/64	.1406	15/64	.2344	13/32	.4062	59/64	.9219
55	.0520	27	.1440	B	.2380	Z	.4130	15/16	.9375
54	.0550	26	.1470	C	.2420	27/64	.4219	61/64	.9531
53	.0595	25	.1495	D	.2460	7/16	.4375	31/32	.9687
1/16	.0625	24	.1520	1/4	.2500	29/64	.4531	63/64	.9844
52	.0635	23	.1540	F	.2570	15/32	.4687	1	1.000
51	.0670	5/32	.1562	G	.2610	31/64	.4843		

CHARTS AND CONVERSION TABLES

APPENDIX H

STANDARD SCREW THREAD CHART

NOMINAL SIZE AND THREADS PER INCH	SERIES DESIGNATION	EXTERNAL CLASS	ALLOWANCE	MAJOR DIA LIMITS MAXa	MAJOR DIA LIMITS MIN	MAJOR DIA LIMITS MINb	PITCH DIA LIMITS MAX	PITCH DIA LIMITS MIN	PITCH DIA LIMITS TOLERANCE	MINOR DIA	INTERNAL CLASS	MINOR DIA LIMITS MIN	MINOR DIA LIMITS MAX	PITCH DIA LIMITS MIN	PITCH DIA LIMITS MAX	PITCH DIA LIMITS TOLERANCE	MAJOR DIA MIN
1-64	UNC	2A	.0006	.0724	.0686	—	.0623	.0603	.0020	.0532	2B	.0561	.0623	.0629	.0655	.0026	.0730
		3A	.0000	.0730	.0692	—	.0629	.0614	.0015	.0538	3B	.0561	.0623	.0629	.0648	.0119	.0730
2-56	UNC	2A	.0006	.0854	.0813	—	.0738	.0717	.0021	.0635	2B	.0667	.0737	.0744	.0772	.0028	.0860
		3A	.0000	.0860	.0819	—	.0744	.0728	.0016	.0641	3B	.0667	.0737	.0744	.0765	.0021	.0860
3-48	UNC	2A	.0007	.0983	.0938	—	.0848	.0825	.0023	.0727	2B	.0764	.0845	.0855	.0885	.0030	.0990
		3A	.0000	.0990	.0945	—	.0855	.0838	.0017	.0734	3B	.0764	.0845	.0855	.0877	.0022	.0990
4-40	UNC	2A	.0008	.1112	.1061	—	.0950	.0925	.0025	.0805	2B	.0849	.0939	.0958	.0991	.0033	.1120
		3A	.0000	.1120	.1069	—	.0958	.0939	.0019	.0813	3B	.0849	.0939	.0958	.0982	.0024	.1120
5-40	UNC	2A	.0008	.1242	.1191	—	.1080	.1054	.0026	.0935	2B	.0979	.1062	.1088	.1121	.0033	.1250
		3A	.0000	.1250	.1199	—	.1088	.1069	.0019	.0943	3B	.0979	.1062	.1088	.1113	.0025	.1250
6-32	UNC	2A	.0008	.1372	.1312	—	.1169	.1141	.0028	.0989	2B	.104	.114	.1177	.1214	.0037	.1380
		3A	.0000	.1380	.1320	—	.1177	.1156	.0021	.0997	3B	.1040	.1140	.1177	.1204	.0027	.1380
6-40	UNF	2A	.0008	.1372	.1321	—	.1210	.1184	.0026	.1065	2B	.111	.119	.1218	.1252	.0034	.1380
		3A	.0000	.1380	.1329	—	.1218	.1198	.0020	.1073	3B	.1110	.1186	.1218	.1243	.0025	.1380
8-32	UNC	2A	.0009	.1631	.1571	—	.1428	.1399	.0029	.1248	2B	.130	.139	.1437	.1475	.0038	.1640
		3A	.0000	.1640	.1580	—	.1437	.1415	.0022	.1257	3B	.1300	.1389	.1437	.1465	.0028	.1640
8-36	UNF	2A	.0008	.1632	.1577	—	.1452	.1424	.0028	.1291	2B	.134	.142	.1460	.1496	.0036	.1640
		3A	.0000	.1640	.1585	—	.1460	.1439	.0021	.1299	3B	.1340	.1416	.1460	.1487	.0027	.1640
10-24	UNC	2A	.0010	.1890	.1818	—	.1619	.1586	.0033	.1379	2B	.145	.156	.1629	.1672	.0043	.1900
		3A	.0000	.1900	.1828	—	.1629	.1604	.0025	.1389	3B	.1450	.1555	.1629	.1661	.0032	.1900
10-32	UNF	2A	.0009	.1891	.1831	—	.1688	.1658	.0030	.1508	2B	.156	.164	.1697	.1736	.0039	.1900
		3A	.0000	.1900	.1840	—	.1697	.1674	.0023	.1517	3B	.1560	.1641	.1697	.1726	.0029	.1900
12-24	UNC	2A	.0010	.2150	.2078	—	.1879	.1845	.0034	.1639	2B	.171	.181	.1889	.1933	.0044	.2160
		3A	.0000	.2160	.2088	—	.1889	.1863	.0026	.1649	3B	.1710	.1807	.1889	.1922	.0033	.2160
12-28	UNF	2A	.0010	.2150	.2085	—	.1918	.1886	.0032	.1712	2B	.177	.186	.1928	.1970	.0042	.2160
		3A	.0000	.2160	.2095	—	.1928	.1904	.0024	.1722	3B	.1770	.1857	.1928	.1959	.0031	.2160

STANDARD SCREW THREAD CHART (CONTINUED)

NOMINAL SIZE AND THREADS PER INCH	SERIES DESIGNATION	CLASS	ALLOWANCE	EXTERNAL MAJOR DIA LIMITS MAXa	MIN	MINb	PITCH DIA LIMITS MAX	MIN	TOLERANCE	MINOR DIA	CLASS	INTERNAL MINOR DIA LIMITS MIN	MAX	PITCH DIA LIMITS MIN	MAX	TOLERANCE	MAJOR DIA MIN
1/4-20	UNC	1A	.0011	.2489	.2367	—	.2164	.2108	.0056	.1876	1B	.196	.207	.2175	.2248	.0073	.2500
		2A	.0011	.2489	.2408	.2367	.2164	.2127	.0037	.1876	2B	.196	.207	.2175	.2223	.0048	.2500
		3A	.0000	.2500	.2419	—	.2175	.2147	.0028	.1887	3B	.1960	.2067	.2175	.2211	.0036	.2500
1/4-28	UNF	1A	.0010	.2490	.2392	—	.2258	.2208	.0050	.2052	1B	.211	.220	.2268	.2333	.0065	.2500
		2A	.0010	.2490	.2425	—	.2258	.2225	.0033	.2052	2B	.211	.220	.2268	.2311	.0043	.2500
		3A	.0000	.2500	.2435	—	.2268	.2243	.0025	.2062	3B	.2110	.2190	.2268	.2300	.0032	.2500
5/16-18	UNC	1A	.0012	.3113	.2982	—	.2752	.2691	.0061	.2431	1B	.252	.265	.2764	.2843	.0079	.3125
		2A	.0012	.3113	.3026	.2982	.2752	.2712	.0040	.2431	2B	.252	.265	.2764	.2817	.0053	.3125
		3A	.0000	.3125	.3038	—	.2764	.2734	.0030	.2443	3B	.2520	.2630	.2764	.2803	.0039	.3125
5/16-24	UNF	1A	.0011	.3114	.3006	—	.2843	.2788	.0055	.2603	1B	.267	.277	.2854	.2925	.0071	.3125
		2A	.0011	.3114	.3042	—	.2843	.2806	.0037	.2603	2B	.267	.277	.2854	.2902	.0048	.3125
		3A	.0000	.3125	.3053	—	.2854	.2827	.0027	.2614	3B	.2670	.2754	.2854	.2890	.0036	.3125
3/8-16	UNC	1A	.0013	.3737	.3595	—	.3331	.3266	.0065	.2970	1B	.307	.321	.3344	.3429	.0085	.3750
		2A	.0013	.3737	.3643	.3595	.3331	.3287	.0044	.2970	2B	.307	.321	.3344	.3401	.0057	.3750
		3A	.0000	.3750	.3656	—	.3344	.3311	.0033	.2983	3B	.3070	.3182	.3344	.3387	.0043	.3750
3/8-24	UNF	1A	.0011	.3739	.3631	—	.3468	.3411	.0057	.3228	1B	.330	.340	.3479	.3553	.0074	.3750
		2A	.0011	.3739	.3667	—	.3468	.3430	.0038	.3228	2B	.330	.340	.3479	.3528	.0049	.3750
		3A	.0000	.3750	.3678	—	.3479	.3450	.0029	.3239	3B	.3300	.3372	.3479	.3516	.0037	.3750
7/16-14	UNC	1A	.0014	.4361	.4205	—	.3897	.3826	.0071	.3485	1B	.360	.376	.3911	.4003	.0092	.4375
		2A	.0014	.4361	.4258	.4206	.3897	.3850	.0047	.3485	2B	.360	.376	.3911	.3972	.0061	.4375
		3A	.0000	.4375	.4272	—	.3911	.3876	.0035	.3499	3B	.3600	.3717	.3911	.3957	.0046	.4375
7/16-20	UNF	1A	.0013	.4362	.4240	—	.4037	.3975	.0062	.3749	1B	.383	.395	.4050	.4131	.0081	.4375
		2A	.0013	.4362	.4281	—	.4037	.3995	.0042	.3749	2B	.383	.395	.4050	.4104	.0054	.4375
		3A	.0000	.4375	.4294	—	.4050	.4019	.0031	.3762	3B	.3830	.3916	.4050	.4091	.0041	.4375
1/2-13	UNC	1A	.0015	.4985	.4822	—	.4485	.4411	.0074	.4041	1B	.417	.434	.4500	.4597	.0097	.5000
		2A	.0015	.4985	.4876	.4822	.4485	.4435	.0050	.4041	2B	.417	.434	.4500	.4565	.0065	.5000
		3A	.0000	.5000	.4891	—	.4500	.4463	.0037	.4056	3B	.4170	.4284	.4500	.4548	.0048	.5000
1/2-20	UNF	1A	.0013	.4987	.4865	—	.4662	.4598	.0064	.4374	1B	.446	.457	.4675	.4759	.0084	.5000
		2A	.0013	.4987	.4906	—	.4662	.4619	.0043	.4374	2B	.446	.457	.4675	.4731	.0056	.5000
		3A	.0000	.5000	.4918	—	.4675	.4643	.0032	.4387	3B	.4460	.4537	.4675	.4717	.0042	.5000

TRIGONOMETRY CHART
OBLIQUE
TRIANGLE

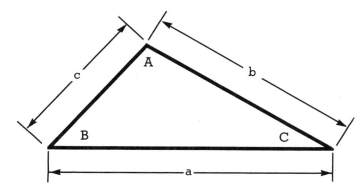

TO FIND	GIVEN	FORMULAS	TO FIND	GIVEN	FORMULAS
A	BC	$180° - (B + C)$	C	AB	$180° - (A + B)$
Tan A	abc	$\dfrac{a \sin C}{b - (a \cos C)}$	Sin C	acA	$\dfrac{c \sin A}{a}$
Cos A	abc	$\dfrac{b^2 + c^2 - a^2}{2bc}$	Tan C	bcA	$\dfrac{c \sin A}{b - (c \cos A)}$
Sin A	acC	$\dfrac{a \sin c}{c}$	Sin C	bcB	$\dfrac{c \sin B}{b}$
Sin A	abB	$\dfrac{a \sin B}{b}$	Tan C	acB	$\dfrac{c \sin B}{a - (c \cos B)}$
Tan A	acB	$\dfrac{a \sin B}{c - (a \cos B)}$	Cos C	abc	$\dfrac{a^2 + b^2 - c^2}{2 ab}$
B	AC	$180° - (A + C)$	c	aAC	$\dfrac{a \sin C}{\sin A}$
Sin B	abA	$\dfrac{b \sin A}{a}$	c	abC	$\sqrt{a^2 + b^2 - (2 ab \cos C)}$
Cos B	abc	$\dfrac{c^2 + a^2 - b^2}{2ac}$	c	bBC	$\dfrac{b \sin C}{\sin B}$
Tan B	bcA	$\dfrac{b \sin A}{c - (b \cos A)}$	AREA	abC	$\dfrac{ab \sin C}{2}$
Sin B	bcC	$\dfrac{b \sin C}{c}$	$s = \dfrac{a + b + c}{2}$	abc	$\sqrt{s\,(s\text{-}a)\,(s\text{-}b)\,(s\text{-}c)}$
a	cAC	$\dfrac{c \sin A}{\sin C}$	b	aAB	$\dfrac{a \sin B}{\sin A}$
a	bAB	$\dfrac{b \sin A}{\sin B}$	b	cBC	$\dfrac{c \sin B}{\sin C}$
a	bcA	$\sqrt{b^2 + c^2 - (2 bc \cos A)}$	b	acB	$\sqrt{a^2 + c^2 - (2 ac \cos B)}$

CHARTS AND CONVERSION TABLES

APPENDIX H

TRIGONOMETRY CHART
RIGHT TRIANGLE

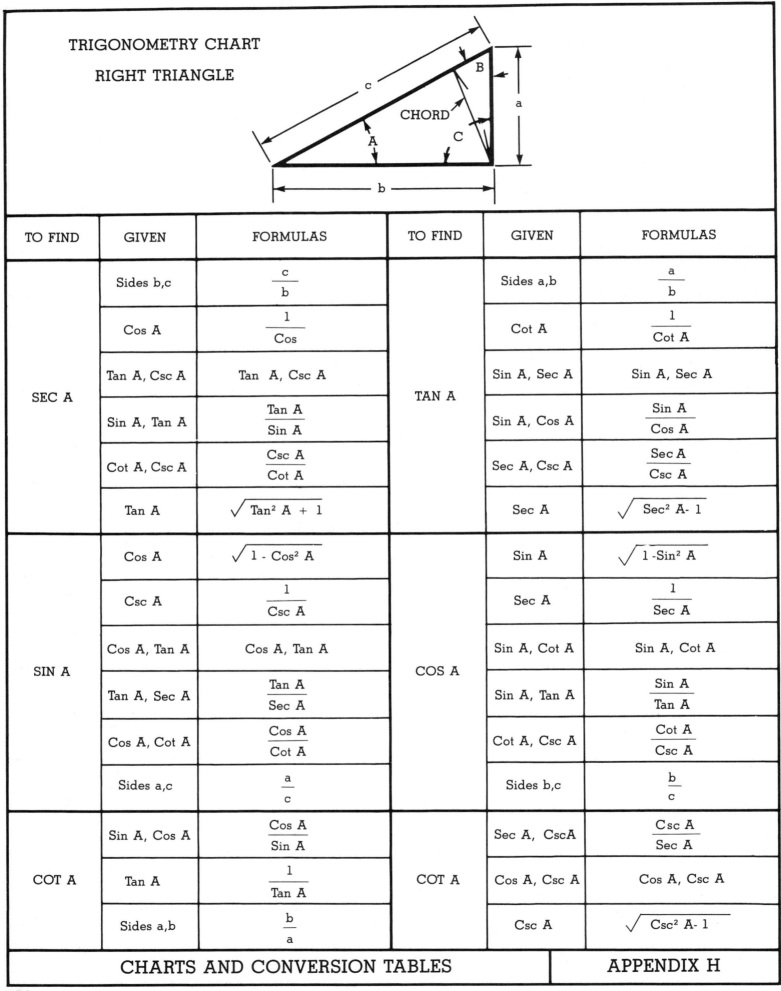

TO FIND	GIVEN	FORMULAS	TO FIND	GIVEN	FORMULAS
SEC A	Sides b,c	$\dfrac{c}{b}$	TAN A	Sides a,b	$\dfrac{a}{b}$
	Cos A	$\dfrac{1}{Cos}$		Cot A	$\dfrac{1}{Cot\ A}$
	Tan A, Csc A	Tan A, Csc A		Sin A, Sec A	Sin A, Sec A
	Sin A, Tan A	$\dfrac{Tan\ A}{Sin\ A}$		Sin A, Cos A	$\dfrac{Sin\ A}{Cos\ A}$
	Cot A, Csc A	$\dfrac{Csc\ A}{Cot\ A}$		Sec A, Csc A	$\dfrac{Sec\ A}{Csc\ A}$
	Tan A	$\sqrt{Tan^2\ A + 1}$		Sec A	$\sqrt{Sec^2\ A - 1}$
SIN A	Cos A	$\sqrt{1 - Cos^2\ A}$	COS A	Sin A	$\sqrt{1 - Sin^2\ A}$
	Csc A	$\dfrac{1}{Csc\ A}$		Sec A	$\dfrac{1}{Sec\ A}$
	Cos A, Tan A	Cos A, Tan A		Sin A, Cot A	Sin A, Cot A
	Tan A, Sec A	$\dfrac{Tan\ A}{Sec\ A}$		Sin A, Tan A	$\dfrac{Sin\ A}{Tan\ A}$
	Cos A, Cot A	$\dfrac{Cos\ A}{Cot\ A}$		Cot A, Csc A	$\dfrac{Cot\ A}{Csc\ A}$
	Sides a,c	$\dfrac{a}{c}$		Sides b,c	$\dfrac{b}{c}$
COT A	Sin A, Cos A	$\dfrac{Cos\ A}{Sin\ A}$	COT A	Sec A, Csc A	$\dfrac{Csc\ A}{Sec\ A}$
	Tan A	$\dfrac{1}{Tan\ A}$		Cos A, Csc A	Cos A, Csc A
	Sides a,b	$\dfrac{b}{a}$		Csc A	$\sqrt{Csc^2\ A - 1}$

CHARTS AND CONVERSION TABLES

APPENDIX H

TRIGONOMETRY CHART
RIGHT TRIANGLE

TO FIND	GIVEN	FORMULAS
COVERS A	Sides a,c	$\dfrac{c-a}{c}$
COVERS A	Sin A	1-Sin A
VERS A	Sides b,c	$\dfrac{c-b}{c}$
VERS A	Cos A	1-Cos A
SIDE b	Sides a,c	$\sqrt{c^2 - a^2}$
SIDE b	Side c, Cos A	c Cos A
SIDE b	Side a, Cot A	a Cot A
SIDE b	Side c, Sec A	$\dfrac{c}{Sec\ A}$
SIDE b	Side a, Tan A	$\dfrac{a}{Tan\ A}$
SIDE a	Sides b,c	$\sqrt{c^2 - b^2}$
SIDE a	Side c, Sin A	c Sin A
SIDE a	Side b, Tan A	b Tan A
SIDE a	Side b, Cot A	$\dfrac{b}{Cot\ A}$
SIDE a	Side c, Csc A	$\dfrac{c}{Csc\ A}$
CHORD	Side b, Sin A	2b Sin $\dfrac{A}{2}$

TO FIND	GIVEN	FORMULAS
AREA	Sides a,b	ab / 2
A	Angles B,C	C-B
B	Angles A,C	C-A
C	Angles A,B	A + B = 90°
CSC A	Sides a,c	$\dfrac{c}{a}$
CSC A	Sin A	$\dfrac{1}{Sin\ A}$
CSC A	Cot A, Sec A	Cot A, Sec A
CSC A	Tan A, Sec A	$\dfrac{Sec\ A}{Tan\ A}$
CSC A	Cos A, Cot A	$\dfrac{Cot\ A}{Cos\ A}$
CSC A	Cot A	$\sqrt{Cot^2\ A + 1}$
SIDE c	Sides a,b	$\sqrt{a^2 + b^2}$
SIDE c	Side b, Sec A	b Sec A
SIDE c	Side a, Csc A	a Csc A
SIDE c	Side a, Sin A	$\dfrac{a}{Sin\ A}$
SIDE c	Side b, Cos A	$\dfrac{b}{Cos\ A}$
SIDE c	Side b, Vers A	$\dfrac{b}{1-Vers\ A}$
SIDE c	Side a, Covers A	$\dfrac{a}{1-Covers\ A}$

MISCELLANEOUS CONVERSION TABLES

MULTIPLY	BY	TO OBTAIN
LENGTH		
MILS	.001	INCHES
MILS	.02540	MILLIMETERS
INCHES	1,000.	MILS
INCHES	25.40	MILLIMETERS
INCHES	2.540	CENTIMETERS
INCHES	.02540	METERS
FEET	30.46	CENTIMETERS
FEET	.3046	METERS
FEET (THOUSANDS OF)	.3046	KILOMETERS
YARDS	.9144	METERS
MILES	1.6093	KILOMETERS
MILLIMETERS	39.37	MILS
MILLIMETERS	.03937	INCHES
CENTIMETERS	.3937	INCHES
CENTIMETERS	.03281	FEET
METERS	39.37	INCHES
METERS	3.281	FEET
METERS	1.0936	YARDS
KILOMETERS	3.281	THOUSANDS OF FEET
KILOMETERS	.6214	MILES
VOLUME		
CUBIC INCHES	16.387	CUBIC CENTIMETERS
CUBIC FEET	.02832	CUBIC METERS
CUBIC CENTIMETERS	.06102	CUBIC INCHES
CUBIC METERS	36.31	CUBIC FEET
QUARTS (LIQUID)	.9464	LITERS
LITERS	1.0567	QUARTS (LIQUID)
GALLONS	231.	CUBIC INCHES
CUBIC INCHES	.004329	GALLONS
MISCELLANEOUS		
POUNDS	.4536	KILOGRAMS
POUNDS PER SQUARE INCH	.07031	KILOGRAMS PER SQUARE CENTIMETER
POUNDS PER CUBIC INCH	27.68	GRAMS PER CUBIC CENTIMETER
POUNDS PER SQUARE INCH	2.03584	INCHES OF MERCURY
GRAMS PER CUBIC CENTIMETER	.03613	POUNDS PER CUBIC INCH
KILOGRAMS	2.2046	POUNDS
KILOGRAMS PER SQUARE CM.	14.223	POUNDS PER SQUARE INCH
INCHES OF MERCURY	.491174	POUNDS PER SQUARE INCH
OHMS PER 1000 FEET	.3281	OHMS PER KILOMETER
OHMS PER KILOMETER	.3048	OHMS PER 1000 FEET

CHARTS AND CONVERSION TABLES	APPENDIX H

MISCELLANEOUS CONVERSION TABLES (CONTINUED)

MULTIPLY	BY	TO OBTAIN
AREA		
SQUARE MILS	1.2732	CIRCULAR MILS
SQUARE MILS	.000001	SQUARE INCHES
CIRCULAR MILS	.7854	SQUARE MILS
CIRCULAR MILS	.0000007854	SQUARE MILS
CIRCULAR MILS	.000001	CIRCULAR INCHES
CIRCULAR MILS	.0005067	SQUARE MILLIMETERS
SQUARE INCHES	1.000.000.	SQUARE MILS
SQUARE INCHES	1,273,200.	CIRCULAR MILS
SQUARE INCHES	1.2732	CIRCULAR INCHES
SQUARE INCHES	645.2	SQUARE MILLIMETERS
SQUARE INCHES	6.452	SQUARE CENTIMETERS
CIRCULAR INCHES	1,000,000.	CIRCULAR MILS
CIRCULAR INCHES	.7854	SQUARE INCHES
SQUARE FEET	.0929	SQUARE METERS
SQUARE MILLIMETERS	1,973.6	CIRCULAR MILS
SQUARE MILLIMETERS	.00155	SQUARE INCHES
SQUARE CENTIMETERS	.15500	SQUARE INCHES
SQUARE METERS	10.764	SQUARE FEET
POWER		
HORSEPOWER	850.	FT-POUNDS PER SECOND
HORSEPOWER	33,000.	FT-POUNDS PER MINUTE
HORSEPOWER	.7457	KILOWATTS
HORSEPOWER	1.014	METRIC HORSEPOWER
FT-POUNDS PER SECOND	.001816	HORSEPOWER
FT-POUNDS PER MINUTE	.00003030	HORSEPOWER
KILOWATTS	1.341	HORSEPOWER
KILOWATTS	1.360	METRIC HORSEPOWER
METRIC HORSEPOWER	.9063	HORSEPOWER
METRIC HORSEPOWER	.7356	KILOWATTS

MENSURATION FORMULAS

TRIANGLE	AREA	= 1/2 (BASE X ALTITUDE)
PARALLELOGRAM	AREA	= BASE X ALTITUDE
TRAPEZOID	AREA	= 1/2 (SUM OF PARALLEL SIDES X ALTITUDE)
CIRCLE	CIRCUMFERENCE	= 3.1416 X DIAMETER = 6.2832 X RADIUS
	AREA	= .07854 X DIAMETER^2 = 3.1416 X RADIUS^2
	AREA OF SECTOR	= 1/2 (RADIUS X ARC)
	AREA OF SEGMENT	= 1/2 (RADIUS X ARC) - 1/2 (RADIUS X CHORD) + 1/2 (CHORD X HEIGHT)
ELLIPSE	AREA	= 0.7854 X SHORT DIAMETER X LONG DIAMETER
CYLINDER	SURFACE	= LENGTH X CIRCUMFERENCE + AREA OF ENDS
	VOLUME	= 0.7854 X LENGTH X DIAMETER^2
CONE	SURFACE (CURVED ONLY)	= 1/2 (SLANT HEIGHT X CIRCUMFERENCE OF BASE)
	VOLUME	= 1/3 (AREA OF BASE X HEIGHT)
	SURFACE	= 3.1416 X DIAMETER^2 = CIRCUMFERENCE X DIAMETER
SPHERE	SURFACE	= 0.5236 X DIAMETER^2 = 1/6 (CIRCUMFERENCE X DIAMETER^2)
	VOLUME	= 2/3 (VOLUME OF CIRCUMSCRIBING CYLINDER)

CHARTS AND CONVERSION TABLES | APPENDIX H

SECTION	EXERCISE	OBJECTIVE: Complete the lettering exercise. Use "H" or "2H" lead. Follow instructions in parentheses. All lettering to be vertical.
1	A	

ABCDEFGHIJKLMNPQRSTUVWXYZ (3/16" HIGH - REPEAT ONCE)

ABCDEFGHIJKLMNOPQRSTUVWXYZ (1/8" HIGH - REPEAT TWICE)

1 2 3 4 5 6 7 8 9 0 (3/16" HIGH - REPEAT 4 TIMES)

1234567890 (1/8" HIGH - REPEAT 6 TIMES)

$2\frac{3}{4}$ $1\frac{7}{8}$ $5\frac{3}{16}$ $9\frac{1}{32}$ (REPEAT ONCE)

TECHNICAL GRAPHICS IS A METHOD OF COMMUNICATION AND IS THE LANGUAGE OF INDUSTRY. THE PREDOMINATE TYPE OF LETTERING USED ON ELECTRONIC DRAWINGS IS THE SINGLE STROKE, UPPER CASE, COMMERCIAL GOTHIC STYLE.   (1/8" HIGH - REPEAT TWICE)

DRAWN BY		DATE	SCALE	DRAWING TITLE	CHECKED BY
				LETTERING	

171

SECTION 1

EXERCISE B

OBJECTIVE: Draw the line conventions described below. Repeat each four times.

CENTER LINE:

PHANTOM LINE:

HIDDEN LINE:

VISIBLE LINE:

CUTTING PLANE LINE:

DRAWN BY	DATE	SCALE	DRAWING TITLE	LINE CONVENTION	CHECKED BY

OBJECTIVE: Scale the drawing below. Using "H" or "2H" lead draw an isometric free hand sketch of the two-view drawing.

DRAWN BY

DATE

SCALE

DRAWING TITLE   ISOMETRIC SKETCHING

CHECKED BY

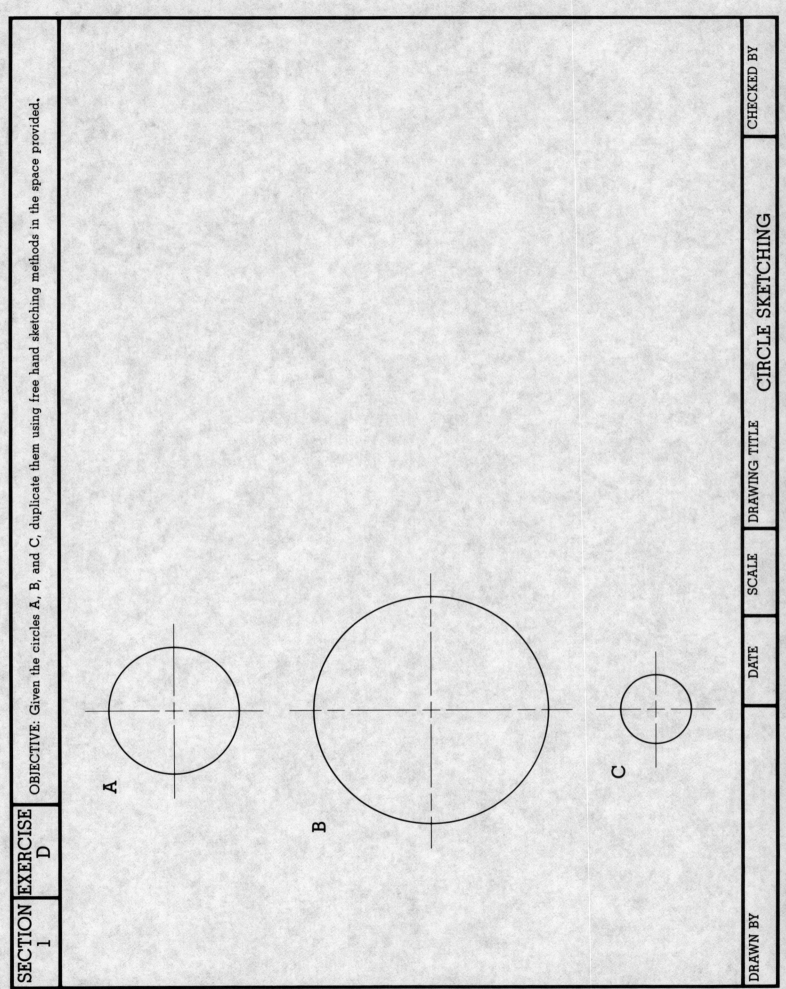

SECTION 1

EXERCISE D

OBJECTIVE: Given the circles A, B, and C, duplicate them using free hand sketching methods in the space provided.

A

B

C

DRAWN BY

DATE | SCALE | DRAWING TITLE   CIRCLE SKETCHING

CHECKED BY

EXERCISE E

OBJECTIVE: Using a scale and drafting tools draw three views of the isometric drawing below (front, top, and right side). Draw full size.

DRAWN BY

DATE

SCALE
FULL

DRAWING TITLE   THREE VIEW DRAWING

CHECKED BY

SECTION 1

EXERCISE F

OBJECTIVE: Scale the two-view drawing below. Add all dimensions and arrow heads, including hole sizes. Use the unidirectional dimensioning method.

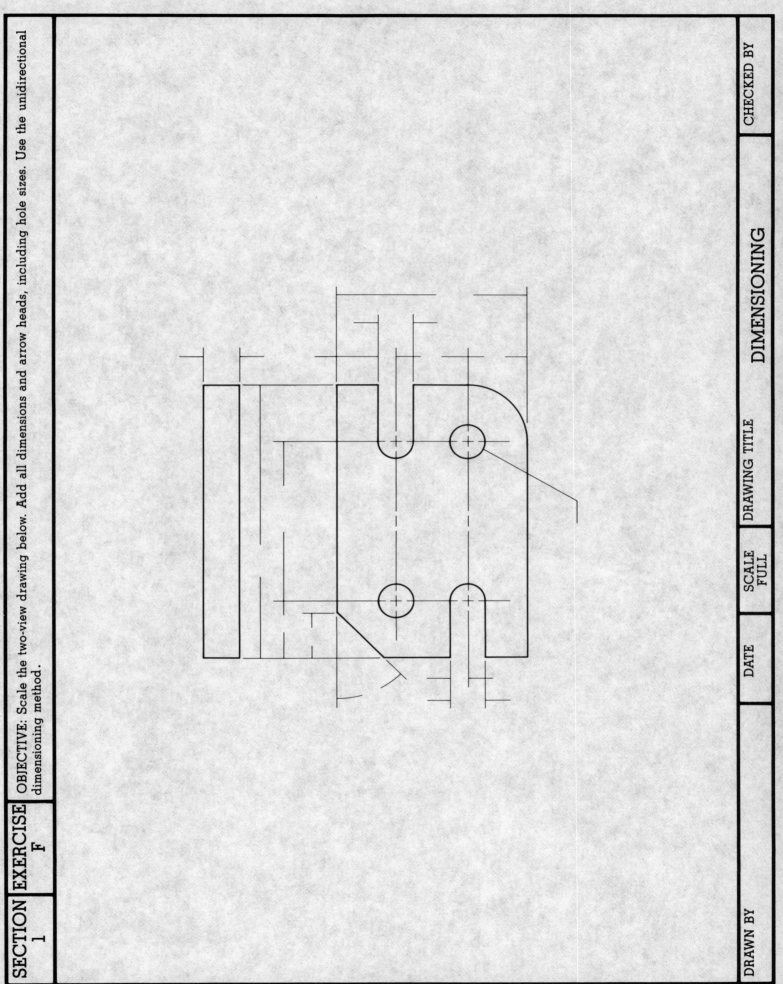

DRAWN BY

DATE

SCALE
FULL

DRAWING TITLE

DIMENSIONING

CHECKED BY

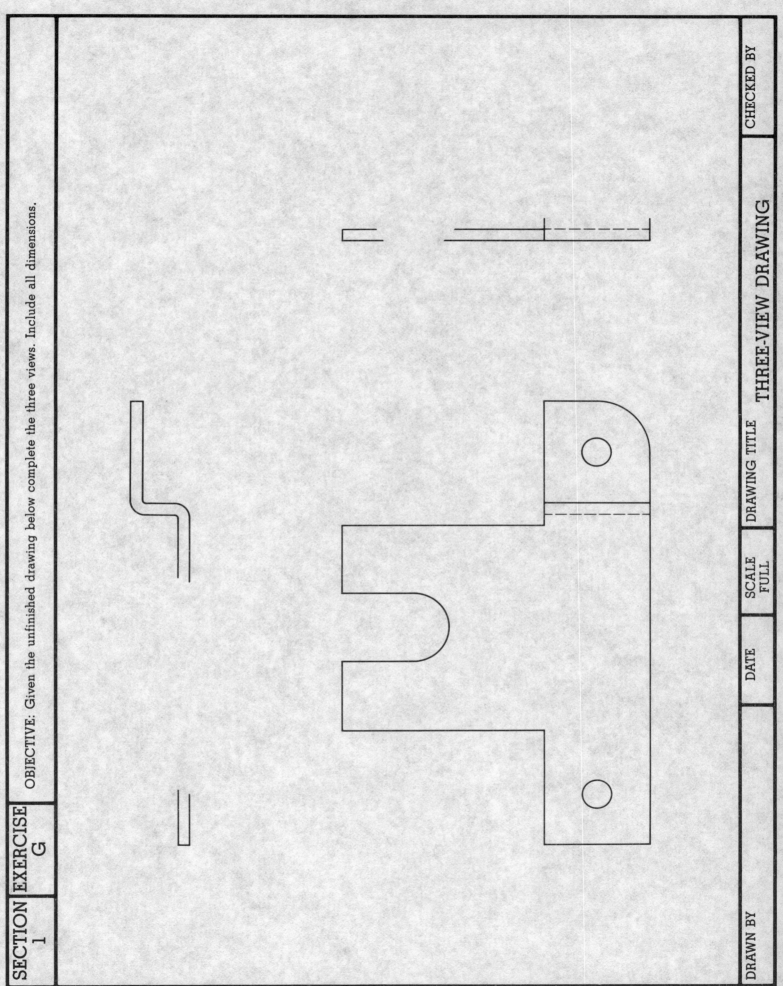

SECTION 1

EXERCISE G

OBJECTIVE: Given the unfinished drawing below complete the three views. Include all dimensions.

DRAWN BY

DATE

SCALE
FULL

DRAWING TITLE

CHECKED BY

THREE-VIEW DRAWING

SECTION 2

EXERCISE A

OBJECTIVE: In the space indicated draw the two most frequently used shapes on a block diagram. Draw four of each to the indicated ratio.

A
1.5:1

B
1:1

DRAWN BY		DATE	SCALE	DRAWING TITLE	CHECKED BY
				BLOCKS	

EXERCISE B

OBJECTIVE: Draw blocks around A, B, and C at a 2:1 ratio. Include flow paths from input to A to B to C to output. Show an auxiliary flow path from the bottom of block A to the bottom of block C. Show flow direction.

INPUT

A

B

C

OUTPUT

DRAWN BY	DATE	SCALE	DRAWING TITLE	BLOCK DIAGRAM	CHECKED BY

187

OBJECTIVE: Complete the block diagram using the information provided. A feeds B, B to C, C to D, D to E, F to A. Add function information to each block. Use word abbreviations from Appendix B.

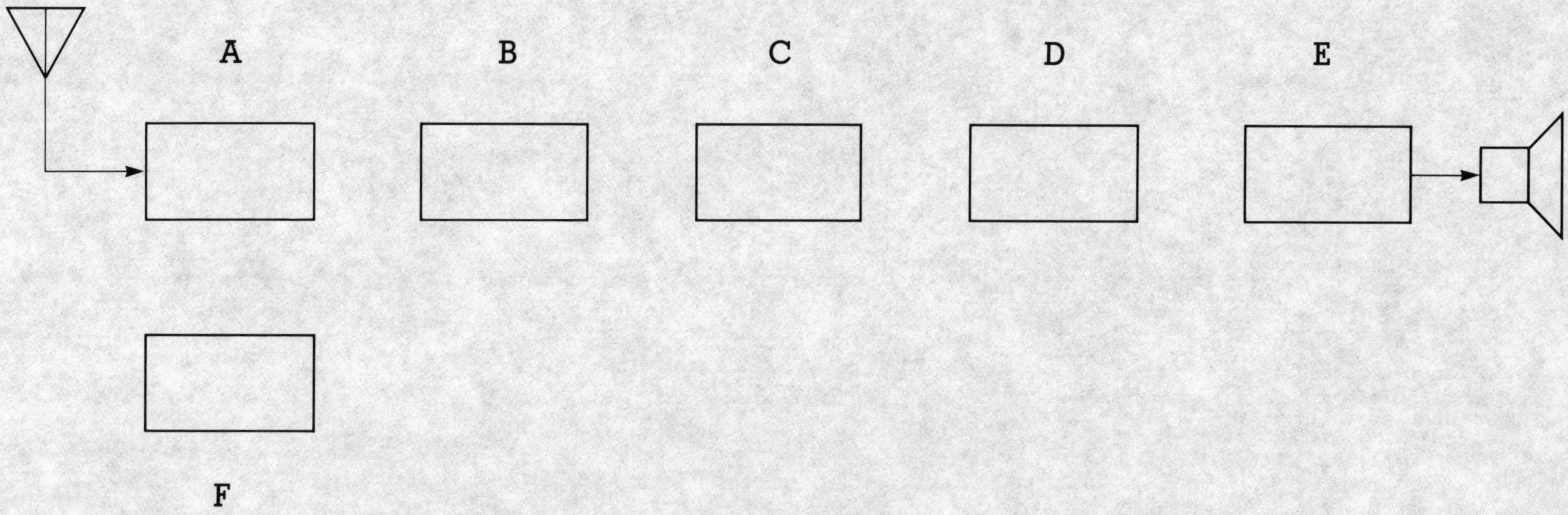

A

B

C

D

E

F

INFORMATION KEY

A = MIXER
B = 1ST INTERMEDIATE FREQUENCY AMPLIFIER
C = 2ND INTERMEDIATE FREQUENCY AMPLIFIER
D = RATIO DETECTOR
E = AUDIO OUTPUT
F = OSCILLATOR

| DRAWN BY | DATE | SCALE | DRAWING TITLE BLOCK DIAGRAM - TRANSISTOR RADIO | CHECKED BY |

OBJECTIVE: Given a free hand sketch of a strip chart recorder circuit, redraw as a block diagram. Add function information to each block.

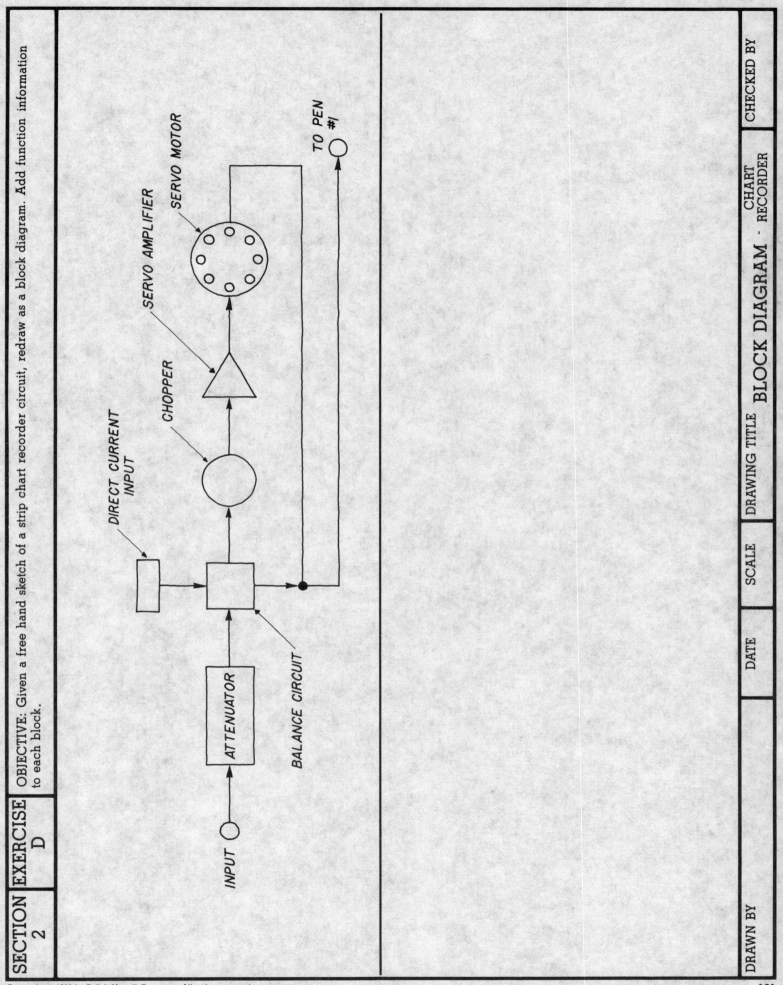

DRAWN BY	DATE	SCALE	DRAWING TITLE	CHECKED BY
			BLOCK DIAGRAM - CHART RECORDER	

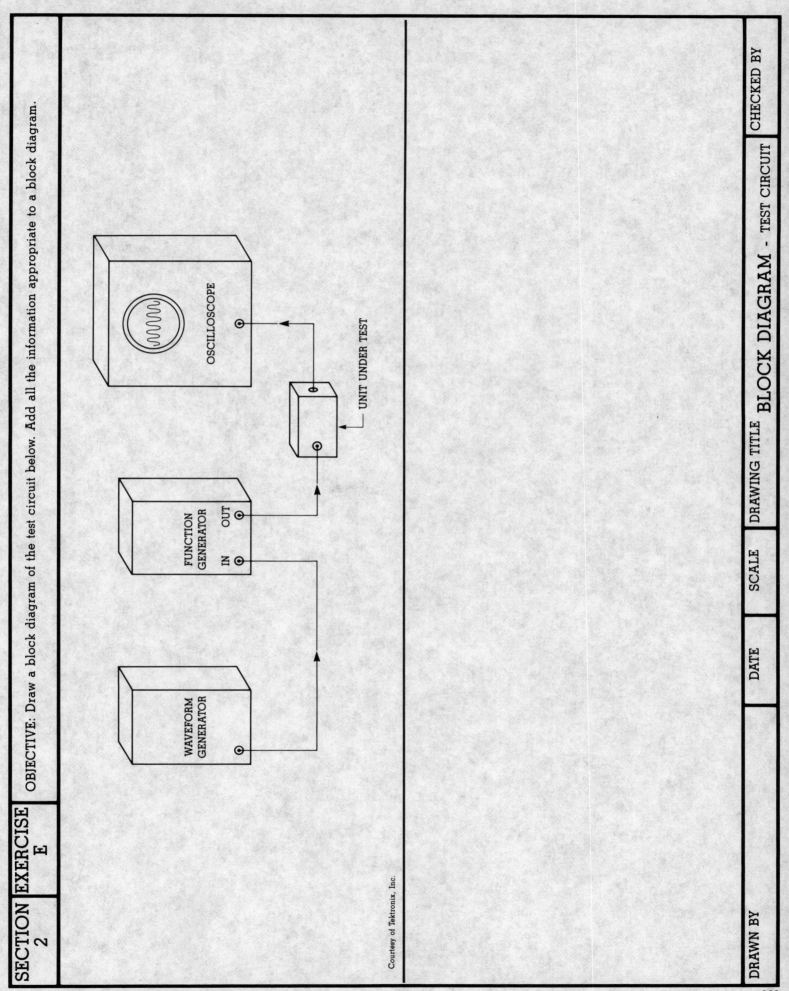

SECTION 2

EXERCISE E

OBJECTIVE: Draw a block diagram of the test circuit below. Add all the information appropriate to a block diagram.

OSCILLOSCOPE

UNIT UNDER TEST

FUNCTION GENERATOR

IN OUT

WAVEFORM GENERATOR

Courtesy of Tektronix, Inc.

DRAWN BY

DATE | SCALE | DRAWING TITLE **BLOCK DIAGRAM** - TEST CIRCUIT

CHECKED BY

SECTION 2	EXERCISE F	OBJECTIVE: Redraw the oscilloscope block diagram shown below using the procedures outlined in this section. All lettering to be upper case and vertical. Draw in the space provided.

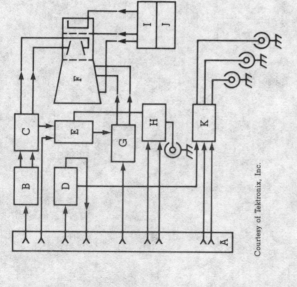

Courtesy of Tektronix, Inc.

INFORMATION KEY

A - Main Frame

B - Vertical Interface

C - Vertical Amplifier

D - Trigger Selector

E - Readout System

F - Cathode Ray Tube

G - Horizontal Amplifier

H - Logic Circuit

I - Cathode Ray Tube Circuit

J - Z-Axis Amplifier

K - Signal Output

HINT: Refer to Appendix B for technical word abbreviations.

DRAWN BY	DATE	SCALE	DRAWING TITLE	CHECKED BY
			BLOCK DIAGRAM, OSCILLISCOPE	

SECTION	EXERCISE
3	A

OBJECTIVE: Shown below are two types of capacitors. Redraw each one twice size in the space provided. Include all information.

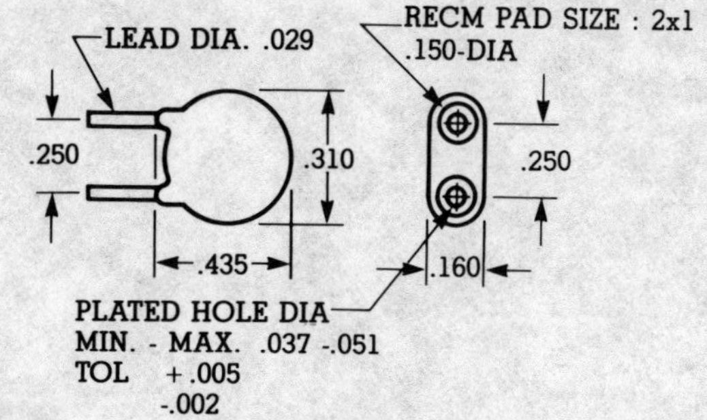

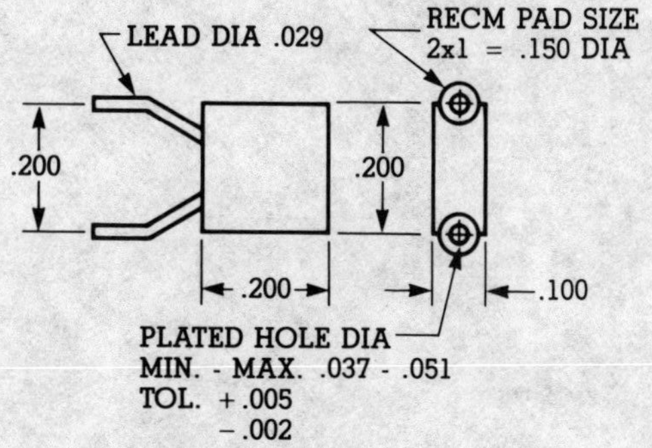

RECM PAD SIZE : 2x1
.150-DIA

.250

.160

LEAD DIA. .029

.310

.435

PLATED HOLE DIA
MIN. - MAX. .037 -.051
TOL. +.005
      -.002

.250

RECM PAD SIZE
2x1 = .150 DIA

.100

.200

LEAD DIA .029

.200

.200

PLATED HOLE DIA
MIN. - MAX. .037 - .051
TOL. +.005
      -.002

.200

DRAWN BY	DATE	SCALE 2:1	DRAWING TITLE COMPONENT DRAWING - CAPACITORS	CHECKED BY

197

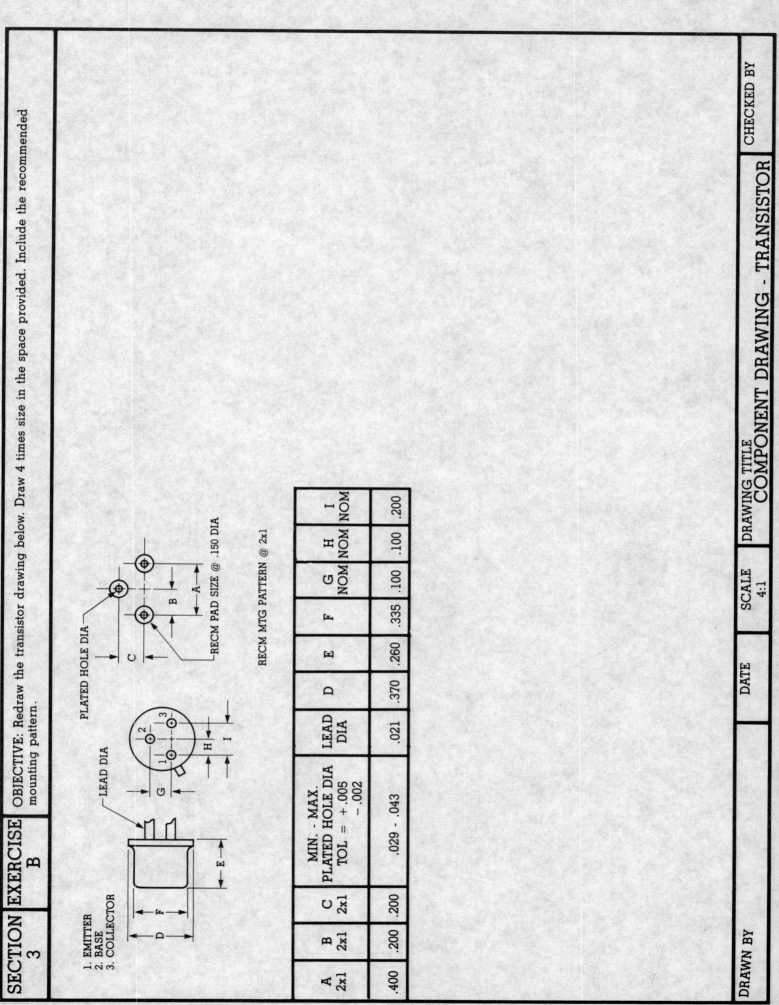

SECTION 3

EXERCISE B

OBJECTIVE: Redraw the transistor drawing below. Draw 4 times size in the space provided. Include the recommended mounting pattern.

1. EMITTER
2. BASE
3. COLLECTOR

PLATED HOLE DIA

LEAD DIA

RECM PAD SIZE @ .150 DIA

RECM MTG PATTERN @ 2x1

A 2x1	B 2x1	C 2x1	MIN. - MAX. PLATED HOLE DIA TOL = +.005 −.002	LEAD DIA	D	E	F	G NOM	H NOM	I NOM
.400	.200	.200	.029 - .043	.021	.370	.260	.335	.100	.100	.200

DRAWN BY

DATE

SCALE 4:1

DRAWING TITLE
COMPONENT DRAWING - TRANSISTOR

CHECKED BY

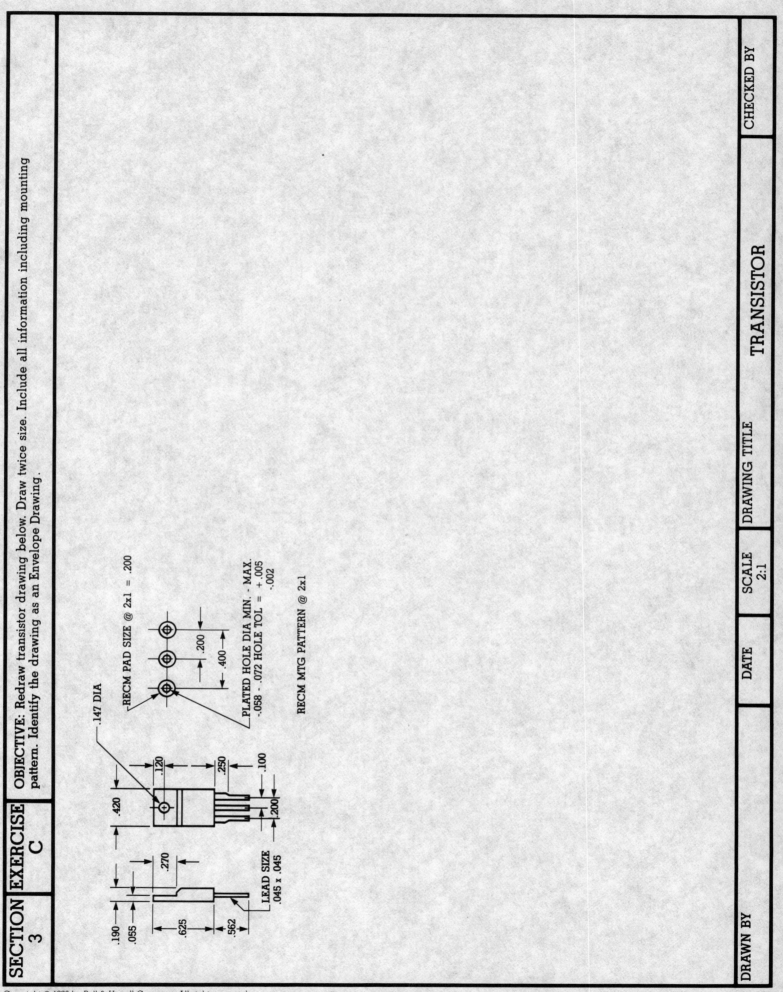

SECTION 3

EXERCISE C

OBJECTIVE: Redraw transistor drawing below. Draw twice size. Include all information including mounting pattern. Identify the drawing as an Envelope Drawing.

.147 DIA

.120

.250

.100

.420

.200

.270

.190

.055

.625

.562

LEAD SIZE
.045 x .045

RECM PAD SIZE @ 2x1 = .200

.200

.400

PLATED HOLE DIA MIN. - MAX.
.058 - .072 HOLE TOL = +.005
                         -.002

RECM MTG PATTERN @ 2x1

DRAWN BY

DATE

SCALE
2:1

DRAWING TITLE

TRANSISTOR

CHECKED BY

SECTION 3

EXERCISE D

OBJECTIVE: In the space provided, redraw twice size the miniature connector identified below. Identify the drawing as a Source Control Drawing.

ONLY THE ITEM DESCRIBED ON THIS DRAWING WHEN PROCURED FROM THE VENDOR(S) LISTED HEREON IS APPROVED BY THE ABC ELECTRONICS CORP. FOR USE IN THE APPLICATION(S) SPECIFIED HEREON. A SUBSTITUTE ITEM SHALL NOT BE USED WITHOUT PRIOR TESTING AND APPROVAL BY ABC ELECTRONICS CORP., ROCHESTER, NY OR BY U.S. ARMY ELECTRONICS MATERIAL SUPPORT AGENCY, FORT MONMOUTH, NEW JERSEY

APPROVED SOURCE(S) OF SUPPLY

VENDOR	VENDOR'S ITEM NO.	APPLICATION
CANNON ELECTRIC CO. 3209 HUMBOLDT ST. LOS ANGELES 31, CALIF.	A-123 SEE NOTE 1	553-MOZ TEST FIXTURE
CODE IDENT NO.:		

NOTE:

1. NUMBER OF CONTACT: 57: 5 AMPS, .054 SPACING (1-57)
   TEST VOLTAGE 60 CPS (AC RMS): 1300 VOLTS

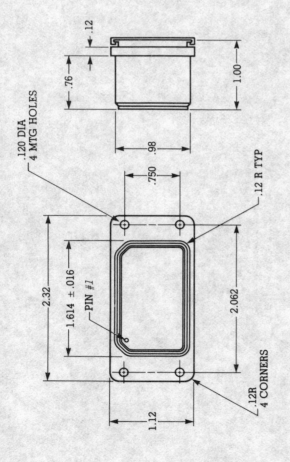

.120 DIA
4 MTG HOLES

.12 R TYP

PIN #1

.12R
4 CORNERS

.12

.76

1.00

.98

.750

2.32

1.614 ± .016

2.062

1.12

DRAWN BY	DATE	SCALE	DRAWING TITLE	CONNECTOR, MINIATURE	CHECKED BY

SECTION 3

EXERCISE E

OBJECTIVE: From Appendix F - Outline Drawings, produce a specification control drawing of the reed relay (K1). Include all performance specifications, vendor information, and schematic. Draw twice size.

DRAWN BY

DATE

SCALE 2:1

DRAWING TITLE RELAY, REED

CHECKED BY

OBJECTIVE: Draw four symbols for each logic function identified below. Use an ANSI Y32.14 logic symbols template or equivalent.

FLIP-FLOP				
EXCLUSIVE OR				
AND				
NOR				
AMPLIFIER				
OR				

DRAWN BY	DATE	SCALE	DRAWING TITLE	LOGIC SYMBOLS
CHECKED BY				

SECTION 4	EXERCISE B	OBJECTIVE: Given the two logic symbols below, add tagging line information to each from the data provided in the information key.

INFORMATION KEY

	AND GATE	NOR GATE
TYPE OF LOGIC FUNCTION		
DRAWING SHEET NUMBER	4	5
CIRCUIT IDENTIFICATION NUMBER	B2A	C3B
MODULE CODE DESIGNATION	G2	G3
REFERENCE DESIGNATION	A5	A7

DRAWN BY	DATE	SCALE	DRAWING TITLE	TAGGING LINES	CHECKED BY

SECTION 4

EXERCISE C

OBJECTIVE: Given a rough sketch of a logic diagram, redraw it in the space provided using a logic symbols template. Add function identification letters to each symbol.

DRAWN BY

DATE | SCALE | DRAWING TITLE | CHECKED BY

LOGIC DIAGRAM

211

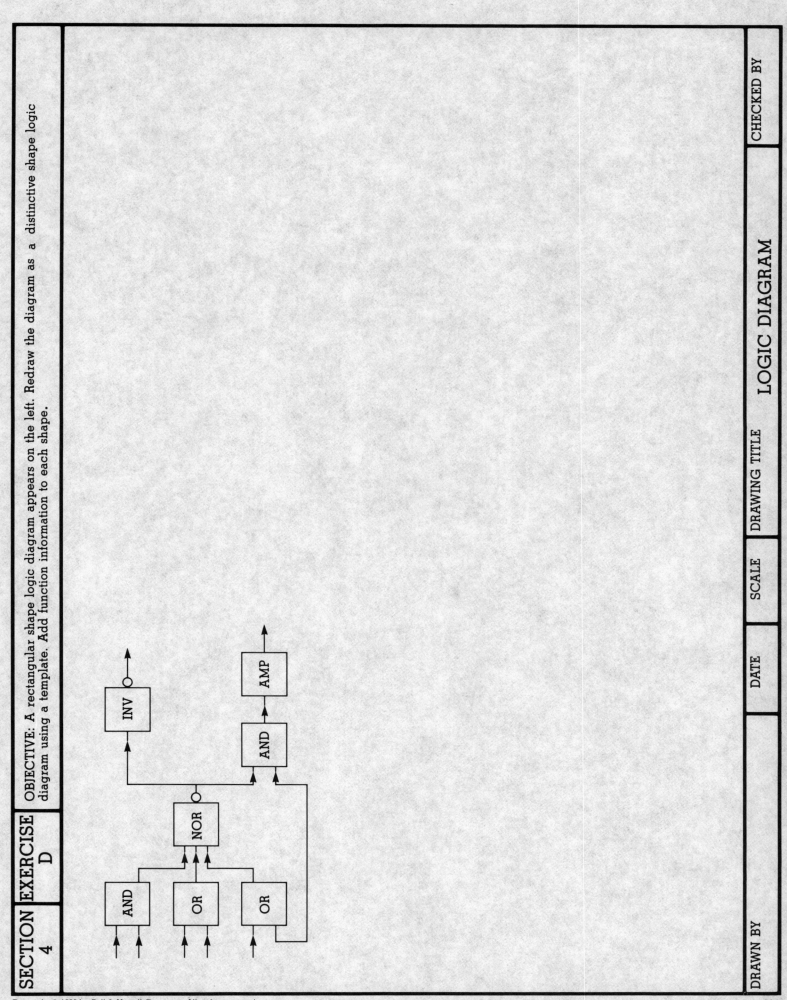

SECTION 4

EXERCISE D

OBJECTIVE: A rectangular shape logic diagram appears on the left. Redraw the diagram as a distinctive shape logic diagram using a template. Add function information to each shape.

DRAWN BY

DATE | SCALE | DRAWING TITLE   LOGIC DIAGRAM

CHECKED BY

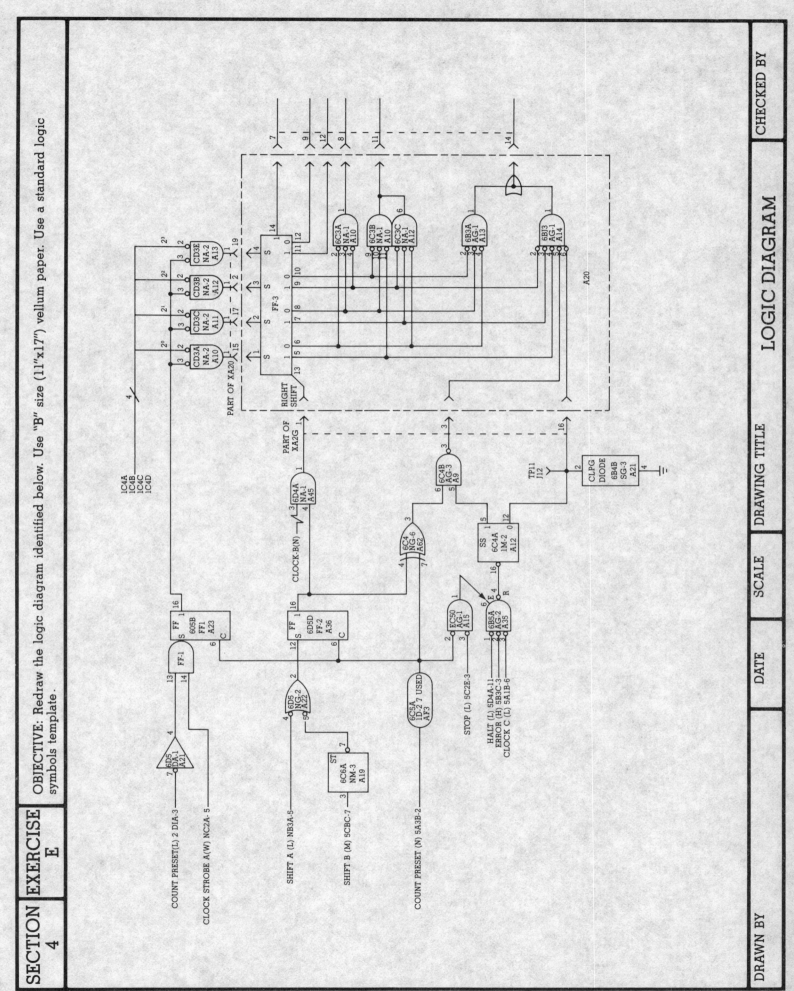

OBJECTIVE: Redraw the logic diagram identified below. Use "B" size (11"x17") vellum paper. Use a standard logic symbols template.

DRAWN BY

DATE

SCALE

DRAWING TITLE

LOGIC DIAGRAM

CHECKED BY

OBJECTIVE: Using an ANSI Y32.2 graphics symbol template or equivalent, draw 19 different electronic symbols with the appropriate class designation letters. Use Appendix A and C for reference.

COMPONENT NAME	CLASS DES. LETTER	GRAPHIC SYMBOL (REPEAT SEVERAL TIMES)	COMPONENT NAME	CLASS DES. LETTER	GRAPHIC SYMBOL
RESISTOR	R	EXAMPLE ⌇⌇⌇⌇			

DRAWN BY	DATE	SCALE	DRAWING TITLE	CHECKED BY
			GRAPHIC SYMBOLS	

OBJECTIVE: Draw 9 different electronic symbols using a template and add both the component sequence number and a component value on each. Use Appendix A and C for reference.

COMPONENT NAME	CLASS DES. LETTER	GRAPHIC SYMBOL (REPEAT SEVERAL TIMES)	COMPONENT NAME	CLASS DES. LETTER	GRAPHIC SYMBOL
CAPACITOR	C	EXAMPLE 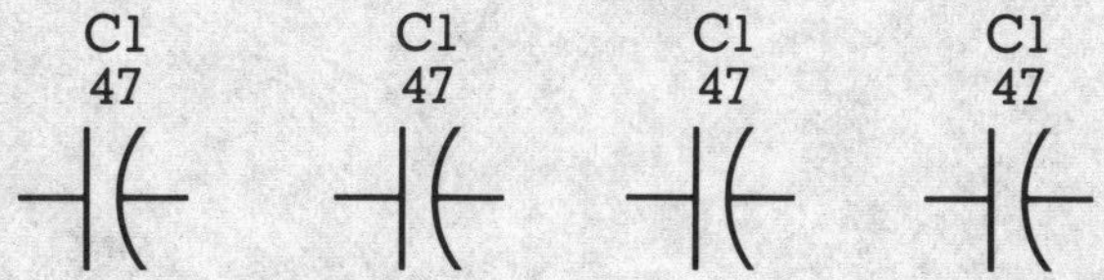 C1 47 C1 47 C1 47 C1 47			

GRAPHIC SYMBOLS

NAME	DATE	SCALE	DRAWING TITLE	CHECKED BY

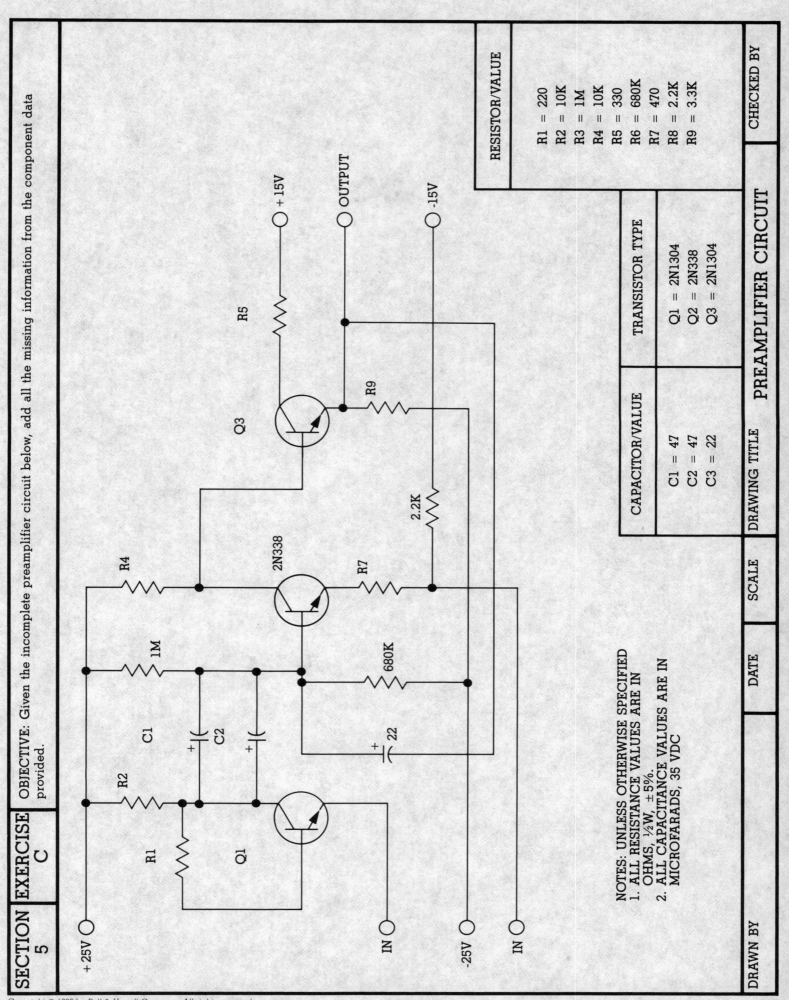

SECTION 5

EXERCISE C

OBJECTIVE: Given the incomplete preamplifier circuit below, add all the missing information from the component data provided.

NOTES: UNLESS OTHERWISE SPECIFIED
1. ALL RESISTANCE VALUES ARE IN OHMS, ½W, ±5%.
2. ALL CAPACITANCE VALUES ARE IN MICROFARADS, 35 VDC

RESISTOR/VALUE
R1 = 220
R2 = 10K
R3 = 1M
R4 = 10K
R5 = 330
R6 = 680K
R7 = 470
R8 = 2.2K
R9 = 3.3K

TRANSISTOR TYPE
Q1 = 2N1304
Q2 = 2N338
Q3 = 2N1304

CAPACITOR/VALUE
C1 = 47
C2 = 47
C3 = 22

DRAWN BY	DATE	SCALE	DRAWING TITLE	PREAMPLIFIER CIRCUIT	CHECKED BY

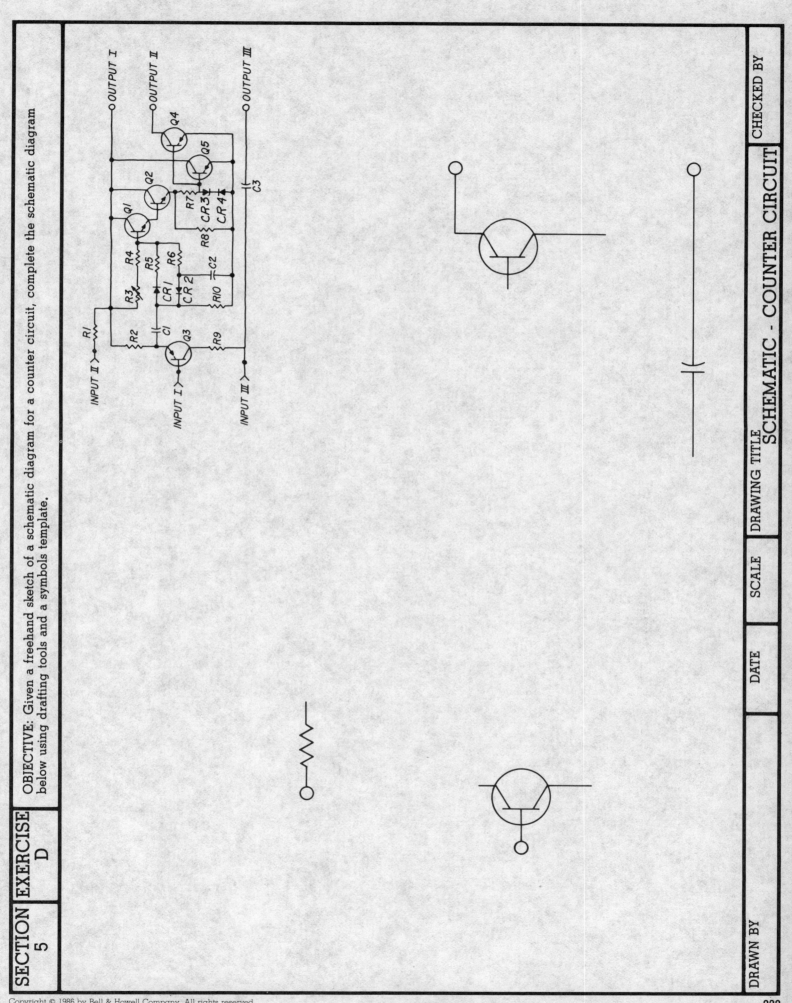

SECTION 5

EXERCISE D

OBJECTIVE: Given a freehand sketch of a schematic diagram for a counter circuit, complete the schematic diagram below using drafting tools and a symbols template.

DRAWN BY | DATE | SCALE | DRAWING TITLE | CHECKED BY

SCHEMATIC - COUNTER CIRCUIT

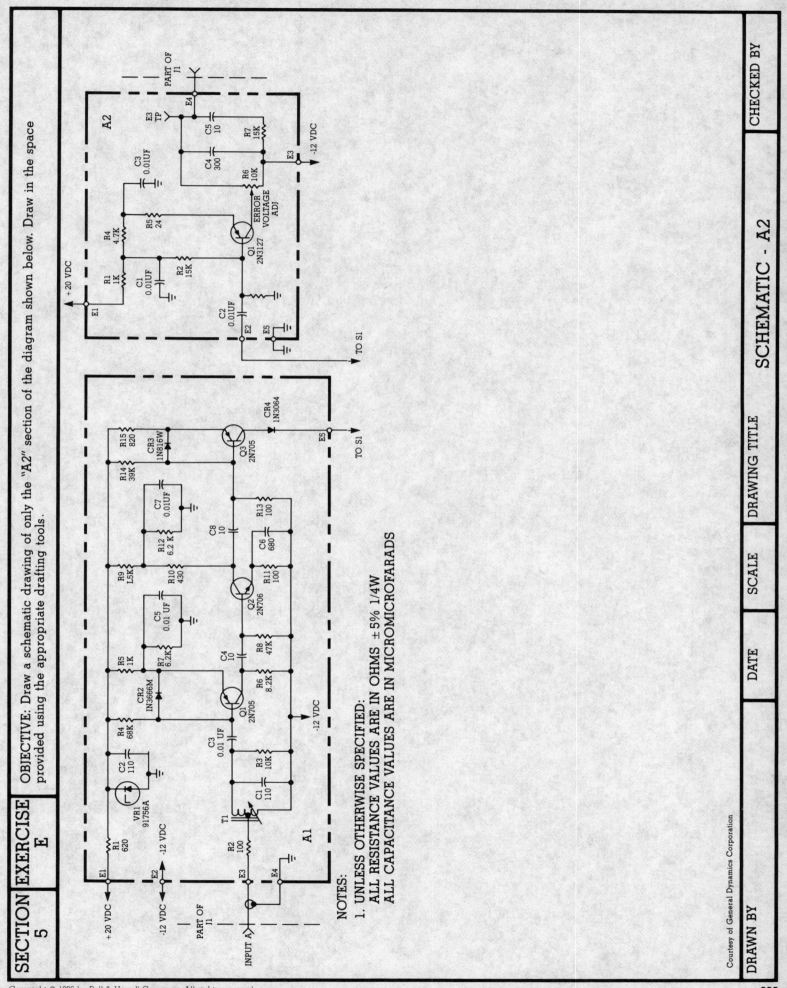

EXERCISE
F

OBJECTIVE: From the schematic diagram identified in the previous exercise (5E) draw the "A1" section in the space provided below.

DRAWN BY

DATE

SCALE

DRAWING TITLE   SCHEMATIC - A1

CHECKED BY

OBJECTIVE: Produce a schematic diagram of the pictorial circuit shown below using drawing practices outlined in this section.

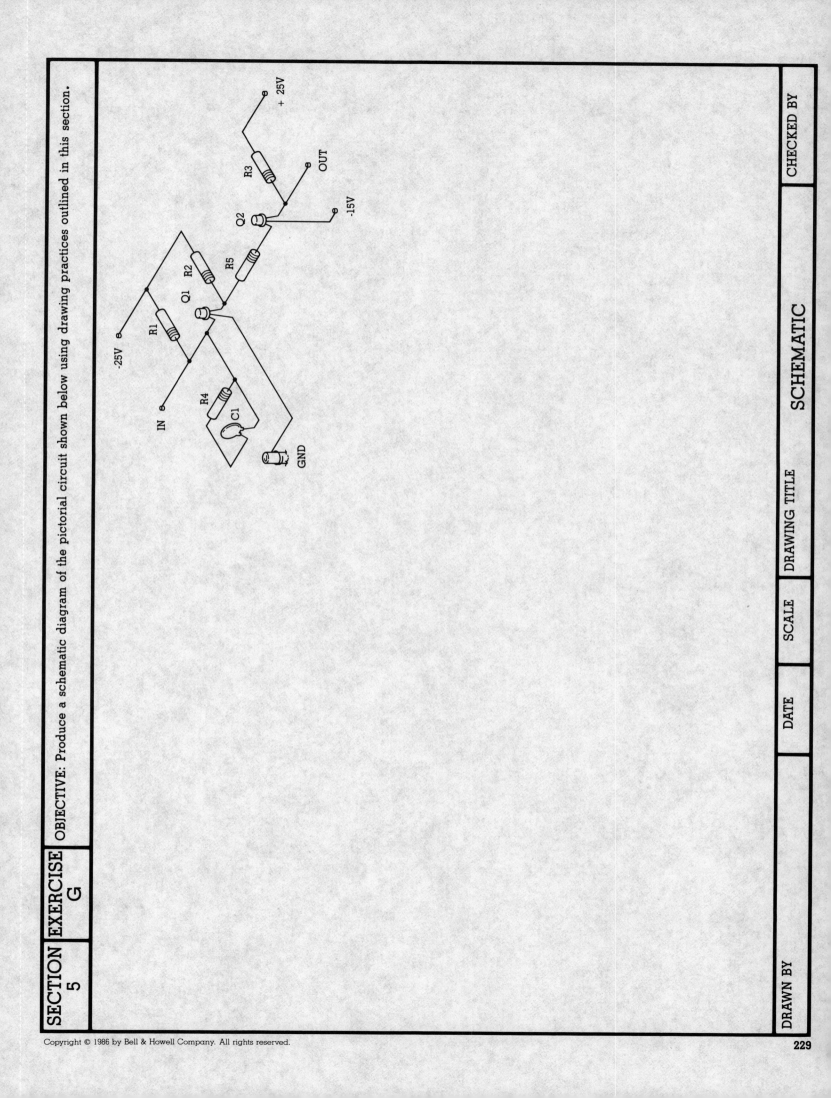

DRAWN BY	DATE	SCALE	DRAWING TITLE	CHECKED BY
			SCHEMATIC	

OBJECTIVE: Redraw the Voltage Sensor schematic shown below. Draw in the space provided.

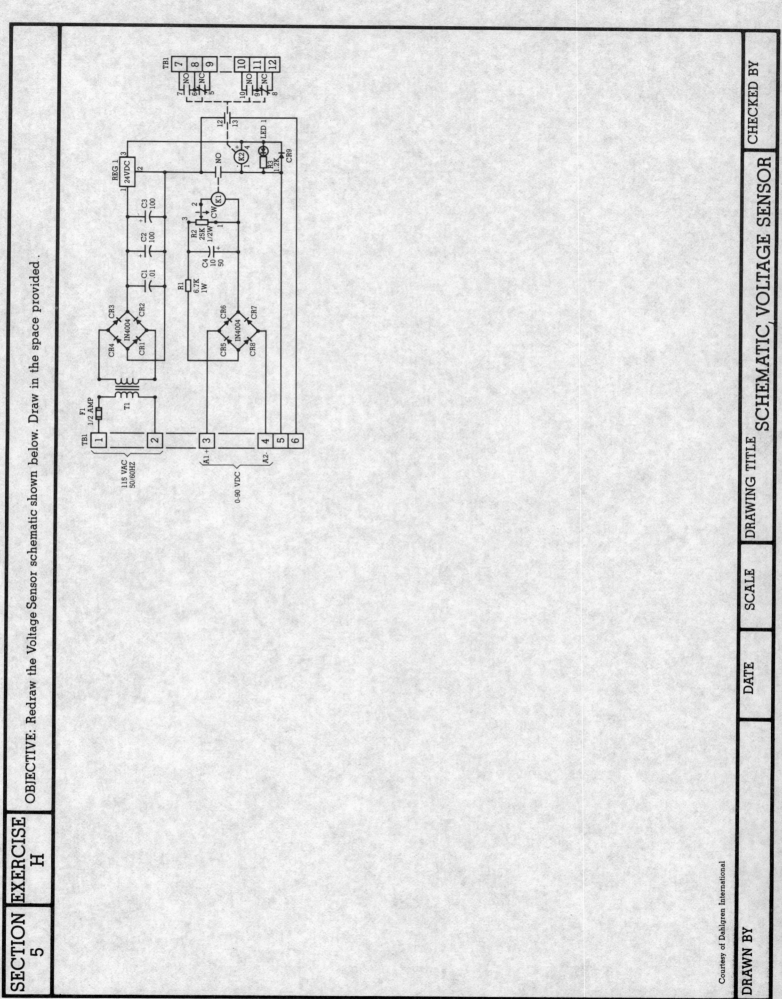

DRAWING TITLE

SCHEMATIC, VOLTAGE SENSOR

DATE | SCALE | DRAWN BY | CHECKED BY

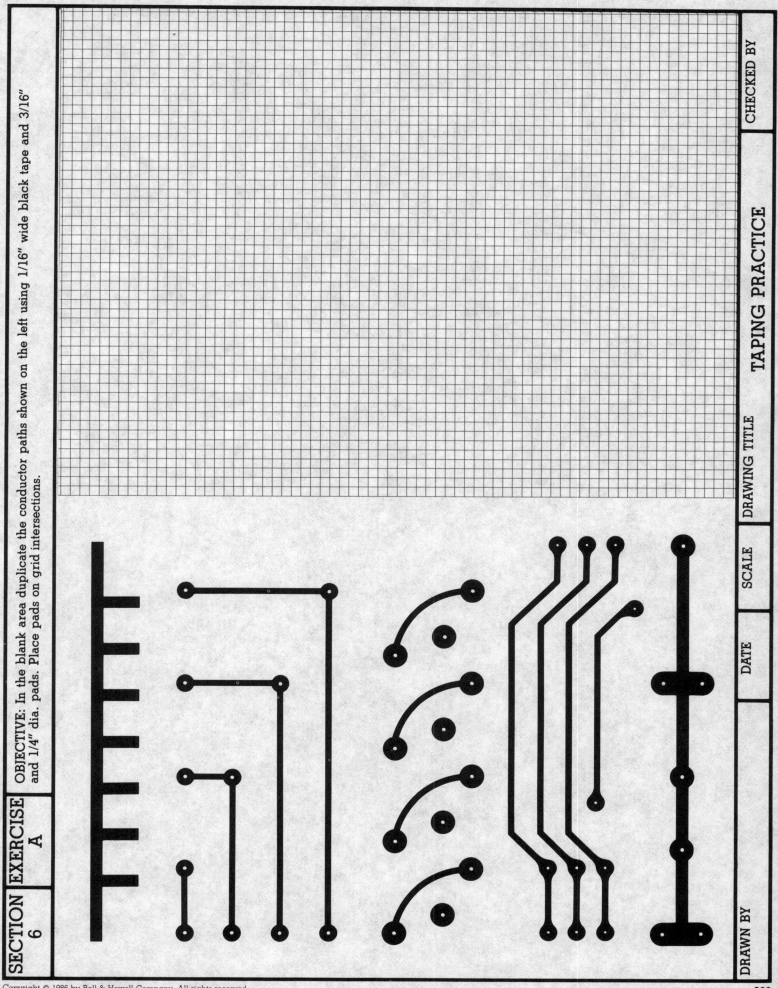

SECTION 6

EXERCISE A

OBJECTIVE: In the blank area duplicate the conductor paths shown on the left using 1/16" wide black tape and 3/16" and 1/4" dia. pads. Place pads on grid intersections.

DRAWN BY

DATE

SCALE

DRAWING TITLE
TAPING PRACTICE

CHECKED BY

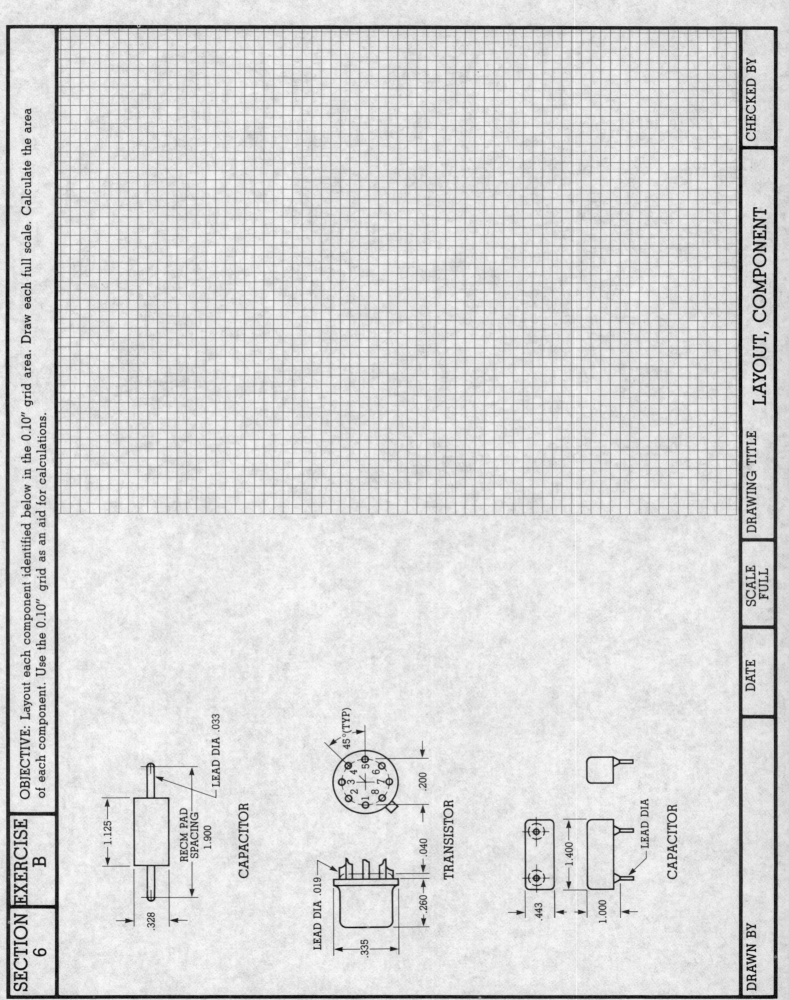

SECTION 6

EXERCISE B

OBJECTIVE: Layout each component identified below in the 0.10″ grid area. Draw each full scale. Calculate the area of each component. Use the 0.10″ grid as an aid for calculations.

CAPACITOR

1.125
RECM PAD SPACING 1.900
LEAD DIA .033
.328

TRANSISTOR

45°(TYP)
.200
.040
.260
LEAD DIA .019
.335

CAPACITOR

.443
1.400
1.000
LEAD DIA

DRAWN BY | DATE | SCALE FULL | DRAWING TITLE LAYOUT, COMPONENT | CHECKED BY

EXERCISE
C

OBJECTIVE: Layout the smallest size component board required to accommodate all the electronic components listed below. Use the 0.10" grid area for the layout. Make layout 2:1 size. Use Appendix G (Bends - Component Leads) for reference.

ELECTRONIC COMPONENTS - FROM APPENDIX F

2 - Diodes - 1 AMP - 400 P.I.V.
3 - Resistors - axial lead - 6.7K OHM - 1W
1 - Variable resistor - 25K OHM - ½W
2 - Capacitors, electrolytic - 10 mfd - 50 VDC

DRAWN BY

DATE

SCALE
2:1

DRAWING TITLE   LAYOUT, COMPONENT BOARD

CHECKED BY

OBJECTIVE: Using 1/16" wide black tape and 1/4" diameter pads, complete the printed circuit pattern by connecting like lettered (A to A, etc.) land areas. Leave 1/16" minimum spacing between conductors. Do not cross conductor paths.

DRAWN BY

DATE | SCALE | DRAWING TITLE | LAYOUT, PRINTED CIRCUIT

CHECKED BY

OBJECTIVE: Using 1/16″ wide black tape connect consecutively lettered land areas. No conductor crossovers are permitted. Leave 1/16″ minimum spacing between conductor paths. (Hint: A1 to A2 to A3 ... C1 to C2, etc.)

DRAWN BY

DRAWING TITLE LAYOUT, PRINTED CIRCUIT

DATE | SCALE | CHECKED BY

SECTION
6

EXERCISE
F

OBJECTIVE: Complete the artwork drawing for PCB-100. Use 1/16" wide black tape to join point 1 with point 1, 2 to 2, 3 to 3, etc. until complete. Allow 1/16" minimum spacing between conductor paths.

DRAWN BY

CHECKED BY

DATE

SCALE

DRAWING TITLE

ARTWORK - PCB-100

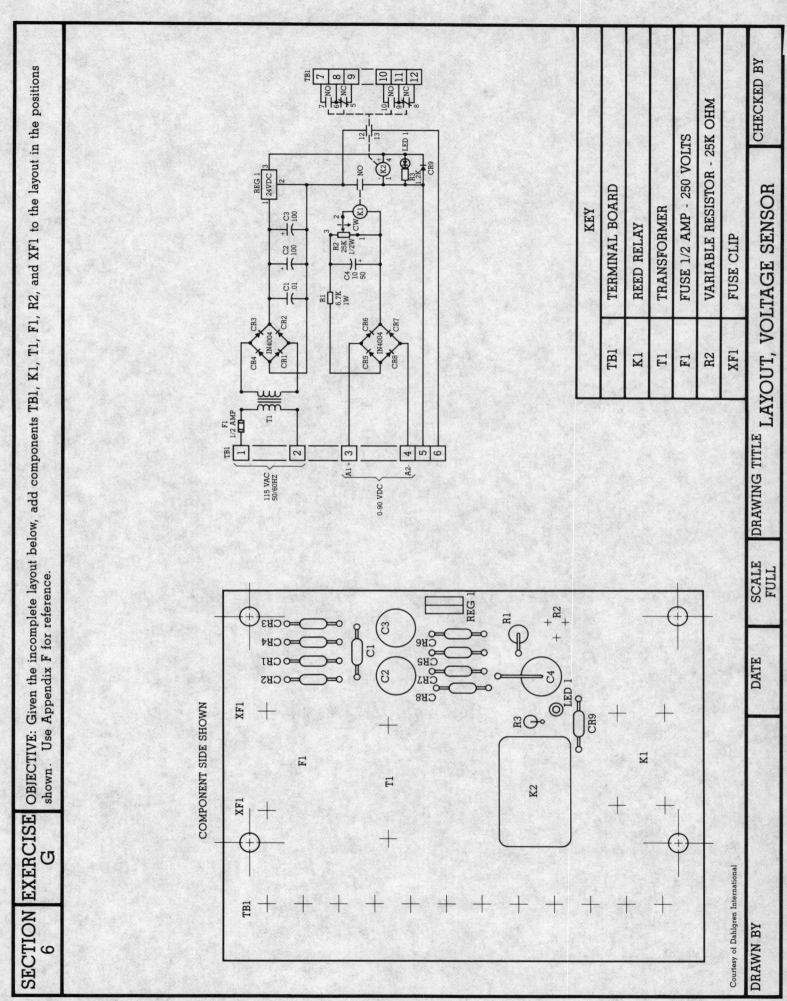

SECTION 6

EXERCISE G

OBJECTIVE: Given the incomplete layout below, add components TB1, K1, T1, F1, R2, and XF1 to the layout in the positions shown. Use Appendix F for reference.

COMPONENT SIDE SHOWN

KEY

TB1	TERMINAL BOARD
K1	REED RELAY
T1	TRANSFORMER
F1	FUSE 1/2 AMP - 250 VOLTS
R2	VARIABLE RESISTOR - 25K OHM
XF1	FUSE CLIP

DRAWN BY	DATE	SCALE FULL	DRAWING TITLE LAYOUT, VOLTAGE SENSOR	CHECKED BY

| SECTION 6 | EXERCISE H | OBJECTIVE: Given partial information produce an artwork drawing at 2:1 scale on 0.10″ grid paper or mylar using 1/8″ wide black tape and 1/4″ and 3/8″ dia. pads. Use the layout below, schematic and layout in exercise 6G for reference. |

NOTE:
BEGIN ARTWORK DRAWING AFTER MAKING CERTAIN THAT ALL THE COMPONENTS CAN BE ACCOMMODATED ON THE LAYOUT.

KEY:

1. ⊕ DENOTES BOARD MOUNTING HOLES - 4 REQ'D

2. ● DENOTES TRANSFORMER (T1) MOUNTING HOLES - 2 REQ'D

3. ▨ REPRESENTS SOLID LAND AREAS

4. —— REPRESENTS CONDUCTOR PATHS FOR WHICH BLACK TAPE (1/8″ WIDE) IS TO BE APPLIED

5. K1, T1, AND TB1 PADS ARE 3/8″ DIA. ALL OTHERS ARE 1/4″ DIA.

CIRCUIT SIDE OF BOARD

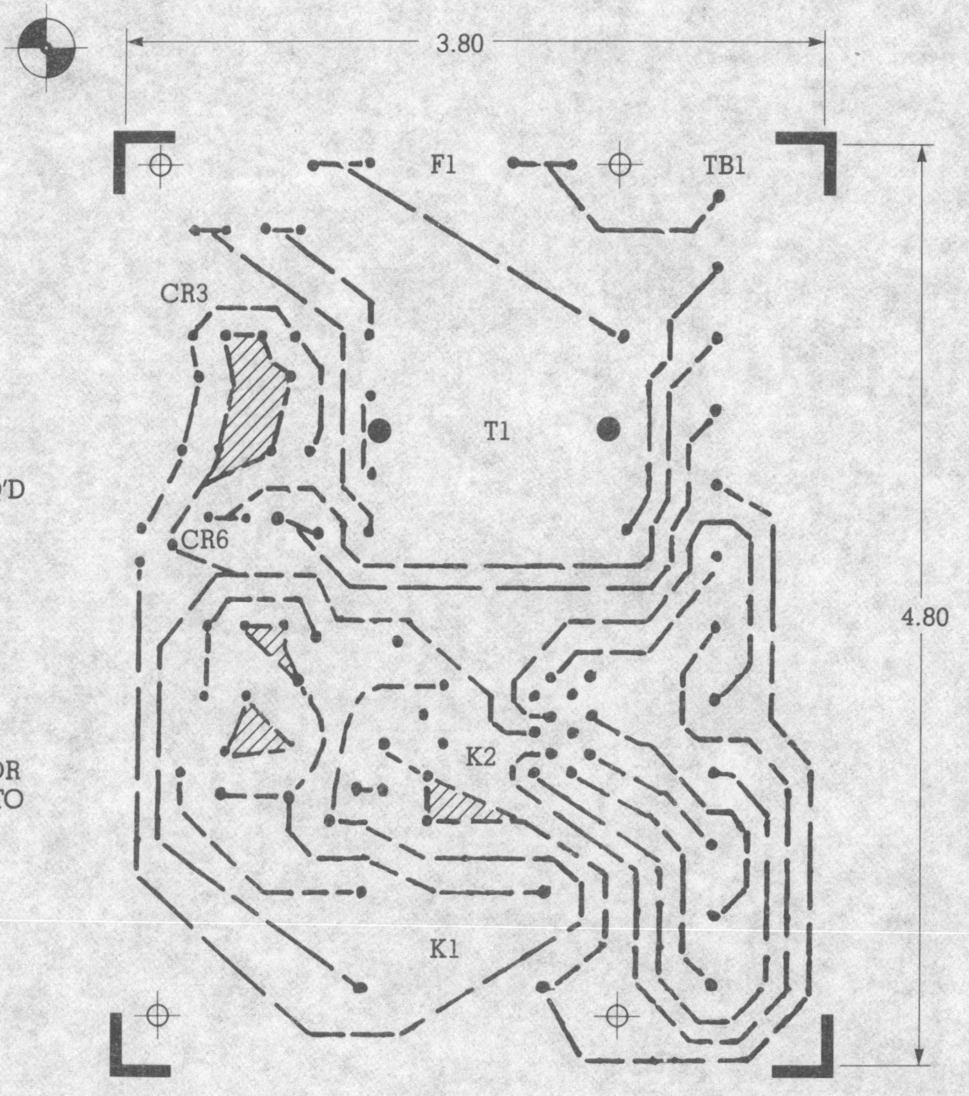

4.80

3.80

TB1

F1

T1

K2

K1

CR3

CR6

SCALE 1:1

| DRAWN BY | | DATE | SCALE | DRAWING TITLE | ARTWORK DRAWING |
| CHECKED BY | | | | | |

247

OBJECTIVE: Complete board detail drawing by adding board outline dimensions, size and location of mounting holes, and all component holes. Add the information to the hole size chart.

HOLE SIZE CHART

SYM	QTY	DRILL	DIA
A			
B			
C			
D			
E			
X			

PRINTED CIRCUIT SIDE SHOWN

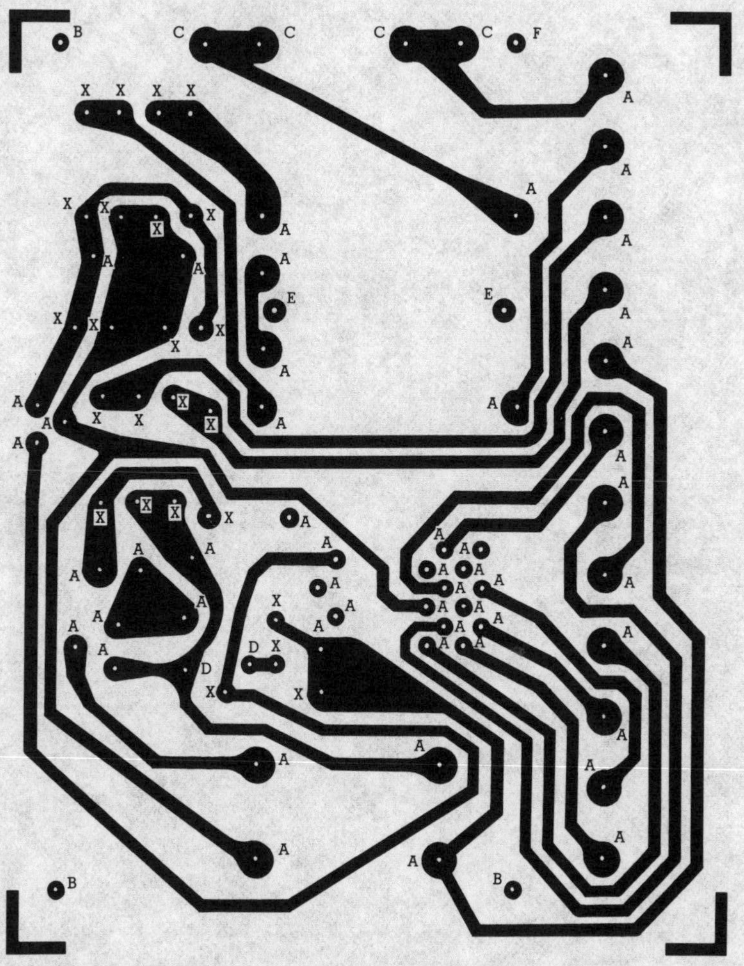

SCALE 1:1

LOCATION OF HOLES CAN
BE OBTAINED FROM 2:1
ARTWORK PRODUCED IN
SECTION 6, EXERCISE "H".

"A" HOLES FOR TB1,K1,K2,CAPACITORS
"B" HOLES - BOARD MTG (.188")
"C" HOLES - FUSE CLIPS
"D" HOLES - LED
"E" HOLES - T1
"X" HOLES - RESISTORS, DIODES

SEE APPENDIX "F" FOR
COMPONENT DATA

DATE	SCALE FULL	DRAWING TITLE DETAIL, BOARD

CHECKED BY

DRAWN BY

Courtesy of Dahlgren International

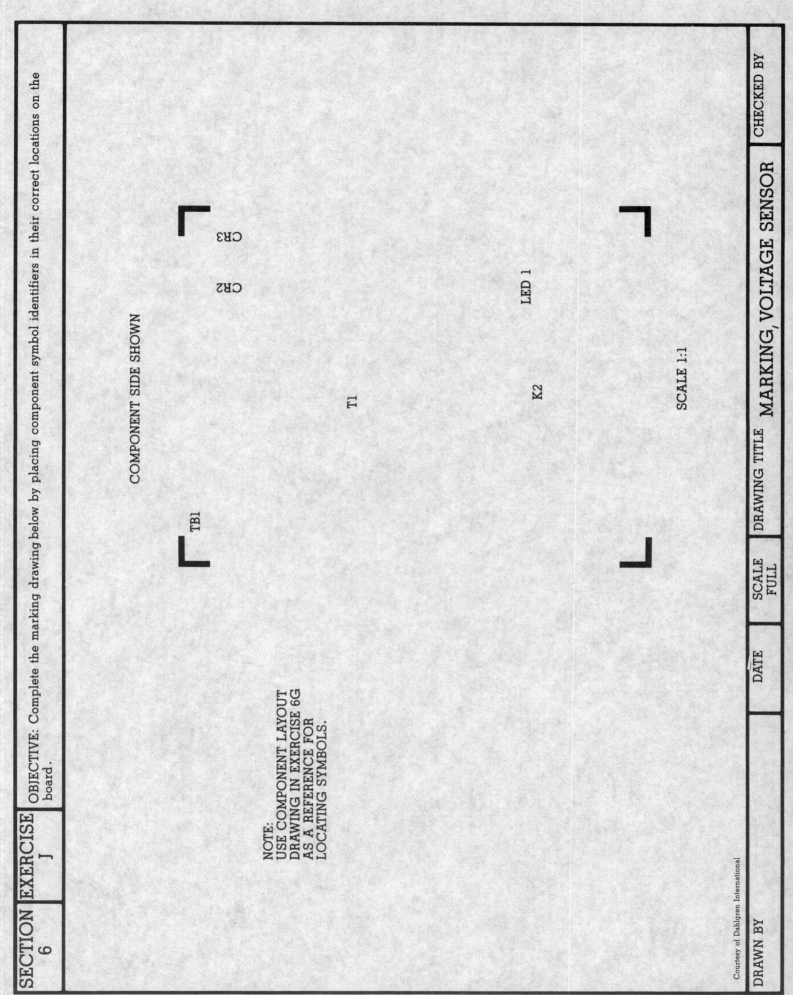

SECTION 6

EXERCISE J

OBJECTIVE: Complete the marking drawing below by placing component symbol identifiers in their correct locations on the board.

COMPONENT SIDE SHOWN

CR3

CR2

T1

TB1

LED 1

K2

SCALE 1:1

NOTE:
USE COMPONENT LAYOUT
DRAWING IN EXERCISE 6G
AS A REFERENCE FOR
LOCATING SYMBOLS.

DATE	SCALE FULL	DRAWING TITLE MARKING, VOLTAGE SENSOR
DRAWN BY		CHECKED BY

251

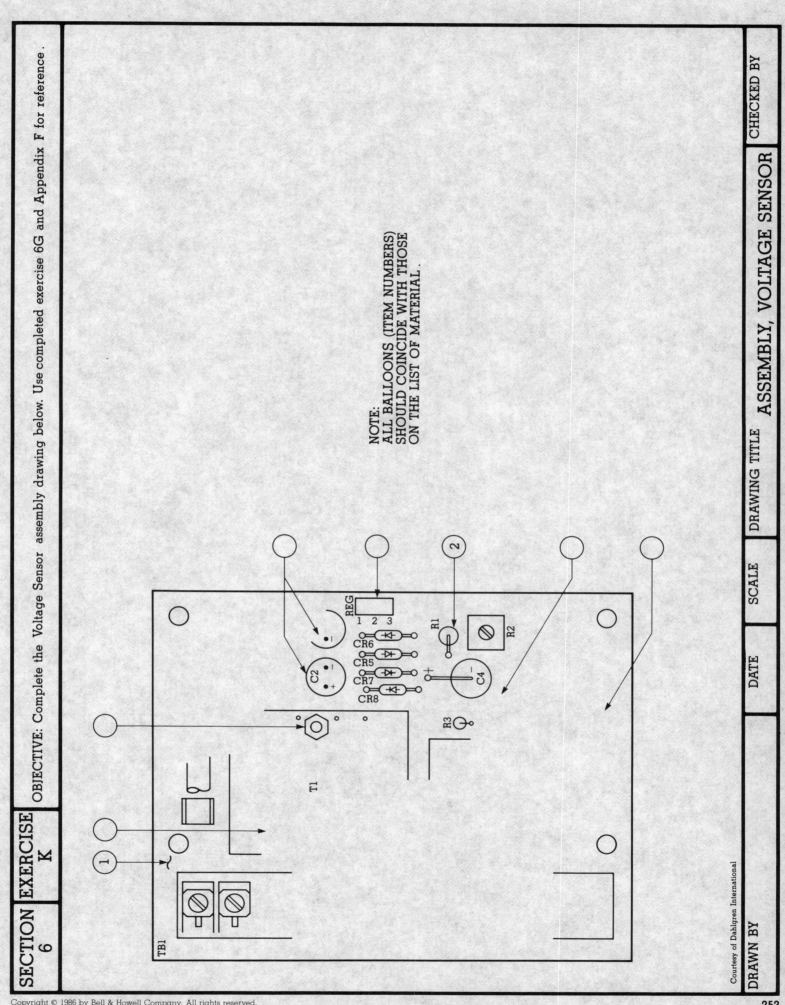

SECTION 6

EXERCISE K

OBJECTIVE: Complete the Voltage Sensor assembly drawing below. Use completed exercise 6G and Appendix F for reference.

NOTE:
ALL BALLOONS (ITEM NUMBERS)
SHOULD COINCIDE WITH THOSE
ON THE LIST OF MATERIAL.

REG
1 2 3

CR6
CR5
CR7
CR8

C2

R1

R2

C4

R3

T1

TB1

① ②

DATE

SCALE

DRAWING TITLE   ASSEMBLY, VOLTAGE SENSOR

CHECKED BY

DRAWN BY

OBJECTIVE: Complete the list of materials below for the Voltage Sensor assembly. All the components identified in Appendix F should be included. Use completed exercise 6K for reference. Assign a 139000-XXX number to each item.

LIST OF MATERIALS

NO.	QTY	SYM	PART NUMBER	DESCRIPTION
1	1	—	139000-001	BOARD, PRINTED CIRCUIT
2	1	R1	139000-002	RESISTOR, 6.7 K OHM, ±5%, 1 WATT
3				
4				
		—		SCREW, PAD HD, CAD. PLT. - 4-40 X 3/8
		—		WASHER, INT. TOOTH, CAD. PLT. - #4
		—		NUT, HEX, CAD. PLT. - 4-40
	2	—		SCREW, PAN. HD. - NYLON - 4-40 X 1 3/8
22	2	—		NUT, HEX - NYLON - 4-40

DRAWN BY	CHECKED BY	DATE	SCALE	DRAWING TITLE L/M, VOLTAGE SENSOR

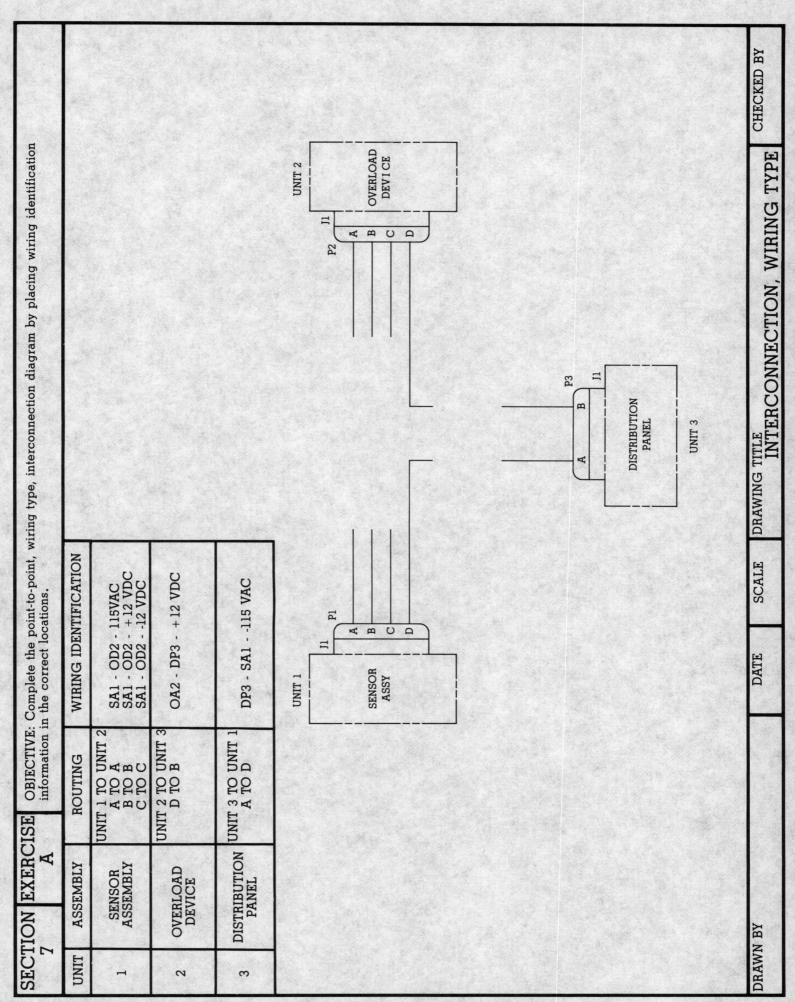

SECTION 7 | EXERCISE A

OBJECTIVE: Complete the point-to-point, wiring type, interconnection diagram by placing wiring identification information in the correct locations.

UNIT	ASSEMBLY	ROUTING	WIRING IDENTIFICATION
1	SENSOR ASSEMBLY	UNIT 1 TO UNIT 2 A TO A B TO B C TO C	SA1 - OD2 - 115VAC SA1 - OD2 - +12 VDC SA1 - OD2 - -12 VDC
2	OVERLOAD DEVICE	UNIT 2 TO UNIT 3 D TO B	OA2 - DP3 - +12 VDC
3	DISTRIBUTION PANEL	UNIT 3 TO UNIT 1 A TO D	DP3 - SA1 - -115 VAC

UNIT 2

P2 J1

A
B
C
D

OVERLOAD DEVICE

UNIT 1

P1 J1

A
B
C
D

SENSOR ASSY

P3 J1

A
B

DISTRIBUTION PANEL

UNIT 3

DRAWN BY | DATE | SCALE | DRAWING TITLE   INTERCONNECTION, WIRING TYPE | CHECKED BY

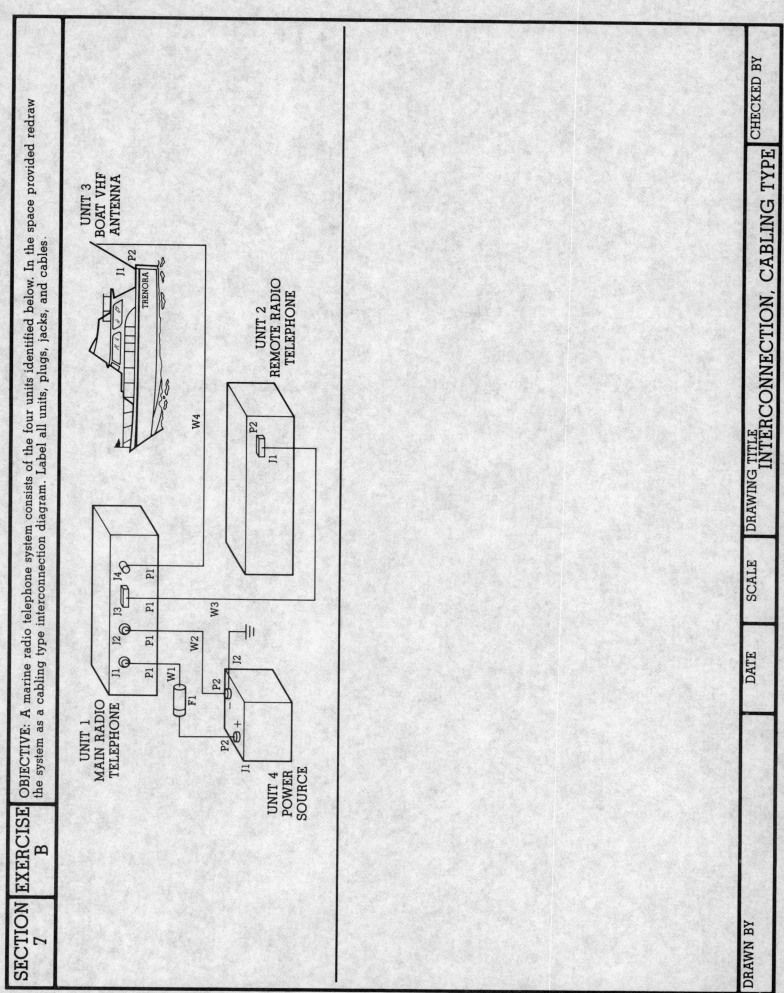

OBJECTIVE: A marine radio telephone system consists of the four units identified below. In the space provided redraw the system as a cabling type interconnection diagram. Label all units, plugs, jacks, and cables.

UNIT 3
BOAT VHF
ANTENNA

P2
J1

TRENORA

W4

UNIT 2
REMOTE RADIO
TELEPHONE

P2
J1

UNIT 1
MAIN RADIO
TELEPHONE

J4 P1
J3 P1
J2 P1
J1 P1

W3

W1 F1
W2
P2
J2

UNIT 4
POWER
SOURCE

P2
J1

DRAWN BY

DATE | SCALE | DRAWING TITLE INTERCONNECTION, CABLING TYPE | CHECKED BY

OBJECTIVE: Redraw the interconnection diagram shown below in the space provided or on B size (11 x 17) vellum.

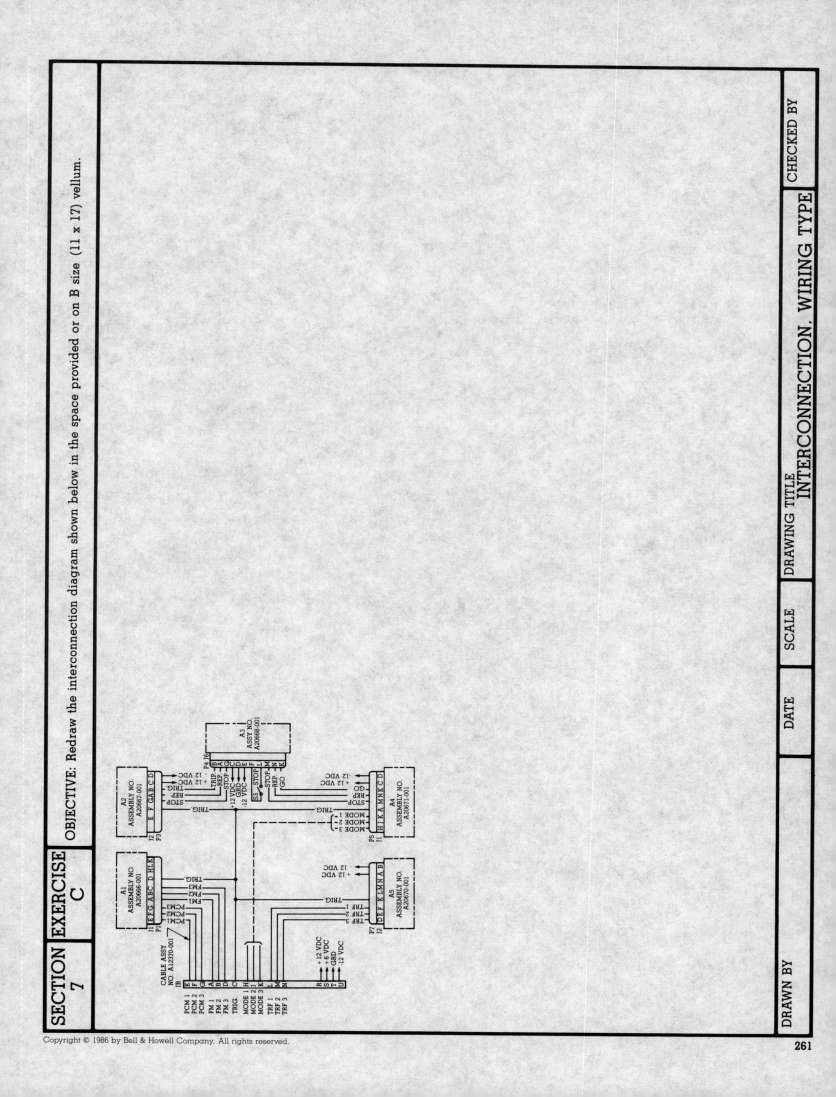

DRAWN BY

DATE | SCALE | DRAWING TITLE | CHECKED BY

INTERCONNECTION. WIRING TYPE

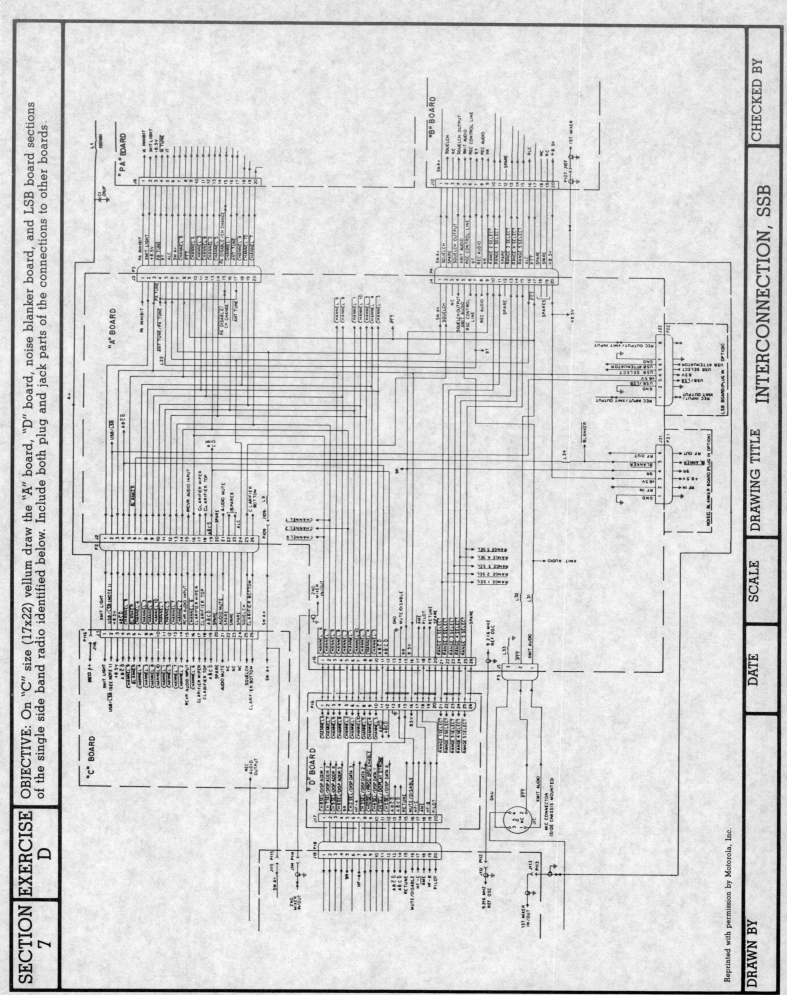

OBJECTIVE: On "C" size (17x22) vellum draw the "A" board, "D" board, noise blanker board, and LSB board sections of the single side band radio identified below. Include both plug and jack parts of the connections to other boards.

CHECKED BY

DRAWING TITLE INTERCONNECTION, SSB

DATE | **SCALE**

DRAWN BY

SECTION 8	EXERCISE A	OBJECTIVE: Each of the symbols below are frequently used on connection diagrams. Draw each five times in the space provided.

TERMINAL BOARD

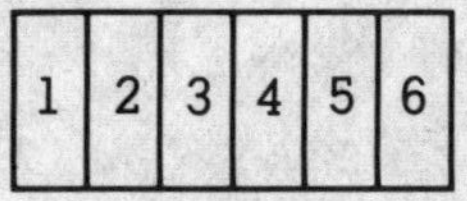

1
2
3
4
5
6

TB4

ROTARY SWITCH

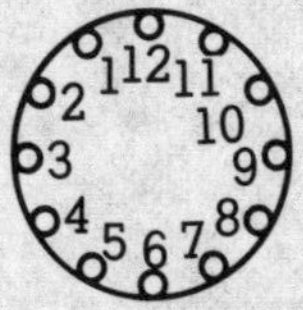

1
2
3
4
5
6
7
8
9
10
11
12

S72

TRANSISTOR

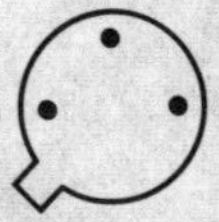

B
E
C

Q3

INTEGRATED CIRCUIT

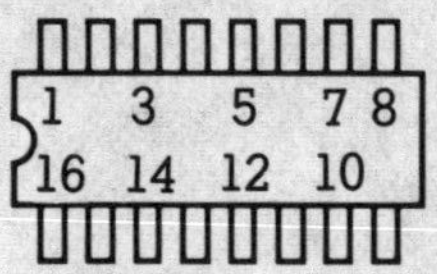

1 3 5 7 8
16 14 12 10

U8

CONNECTOR

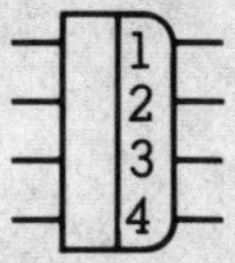

1
2
3
4

J2 P2

DRAWN BY	DATE	SCALE	DRAWING TITLE SYMBOLS, CONNECTION DIAGRAM	CHECKED BY

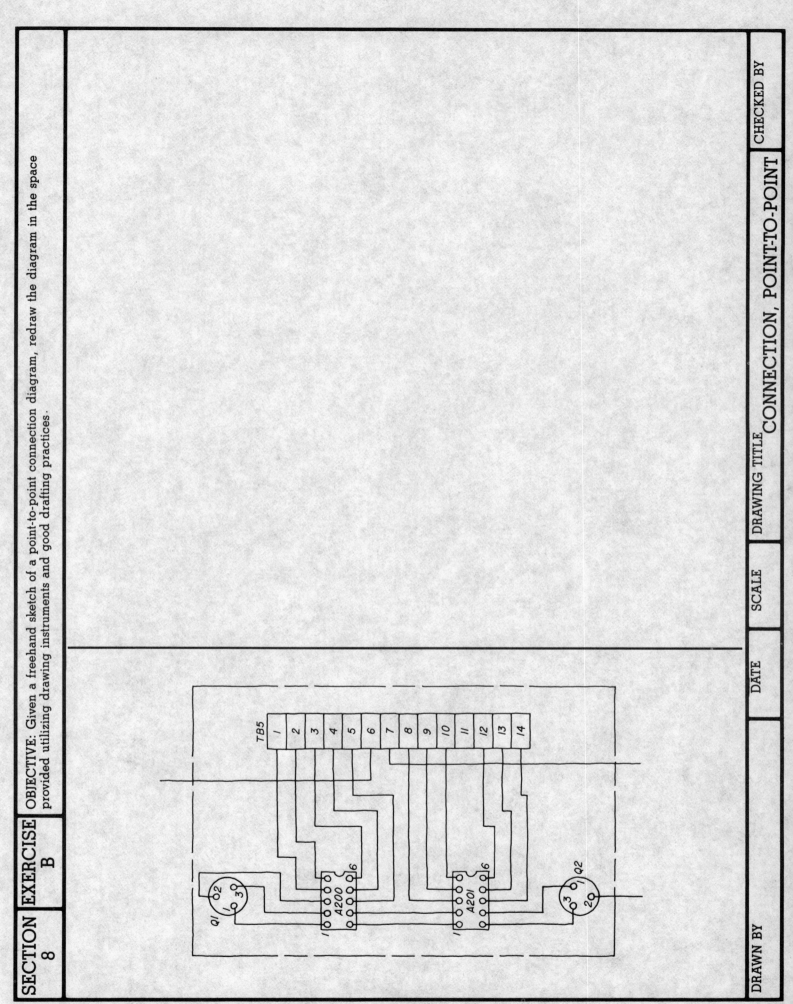

SECTION 8 EXERCISE C

OBJECTIVE: In the space below redraw the point-to-point diagram in exercise 8B (the previous exercise) as a highway type connection diagram.

DRAWN BY

DATE

SCALE

DRAWING TITLE CONNECTION, HIGHWAY TYPE

CHECKED BY

OBJECTIVE: Transfer the data below to the wire data list. Use conductor color code abbreviations from Appendix D.

63000 - WIRE #1 - 14 GAGE - BLUE - CONNECTS A2 WITH B3
63001 - WIRE #2 - 16 GAGE - GREEN - CONNECTS A3 WITH B4
63002 - WIRE #3 - 12 GAGE - RED - CONNECTS A4 WITH B5
63003 - WIRE #4 - 16 GAGE - BLACK - CONNECTS A12 WITH C14
63004 - WIRE #5 - 16 GAGE - ORANGE - CONNECTS A13 WITH C15
63005 - WIRE #6 - 14 GAGE - BROWN - CONNECTS A14 WITH C16
63006 - WIRE #7 - 18 GAGE - WHITE - CONNECTS D4 WITH E6
63007 - WIRE #8 - 18 GAGE - VIOLET - CONNECTS D5 WITH E7
63008 - WIRE #9 - 20 GAGE - PURPLE - CONNECTS F9 WITH G12

WIRE DATA LIST

WIRE NO.	GAGE	COLOR	STOCK NUMBER	FROM	TO

DRAWN BY	DATE	SCALE	DRAWING TITLE	DATA LIST, WIRE	CHECKED BY

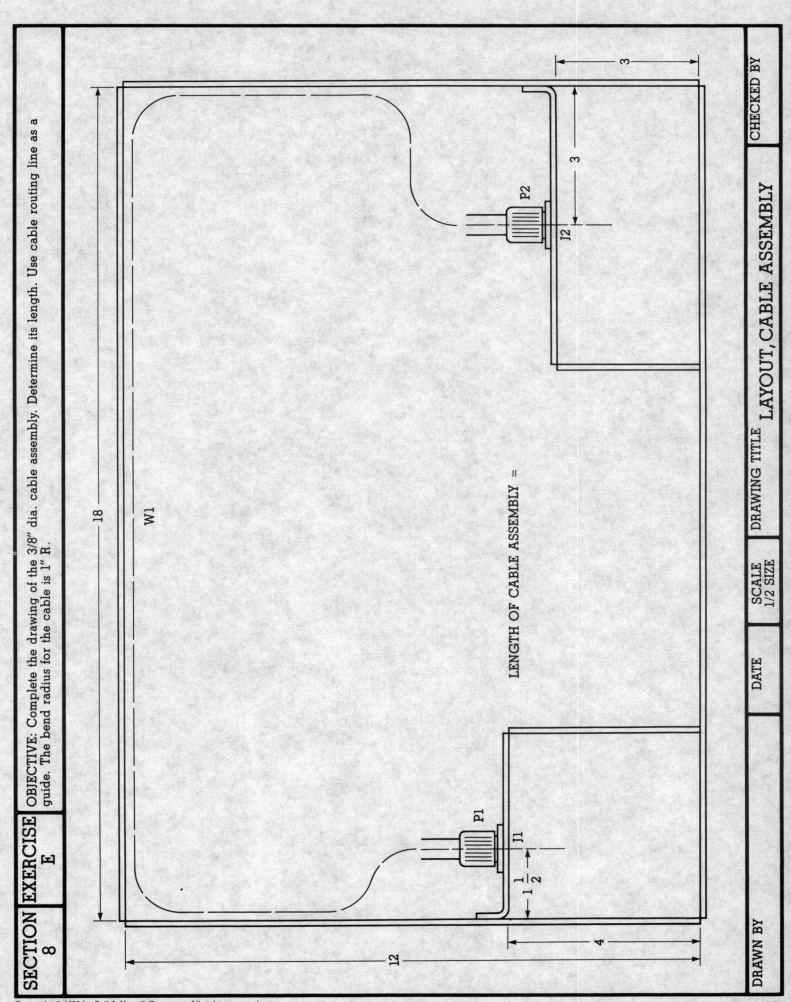

SECTION 8

EXERCISE E

OBJECTIVE: Complete the drawing of the 3/8″ dia. cable assembly. Determine its length. Use cable routing line as a guide. The bend radius for the cable is 1″ R.

18

W1

3

3

J2

P2

1 1/2

1

J1

P1

4

12

LENGTH OF CABLE ASSEMBLY =

DRAWN BY

DATE

SCALE
1/2 SIZE

DRAWING TITLE   LAYOUT, CABLE ASSEMBLY

CHECKED BY

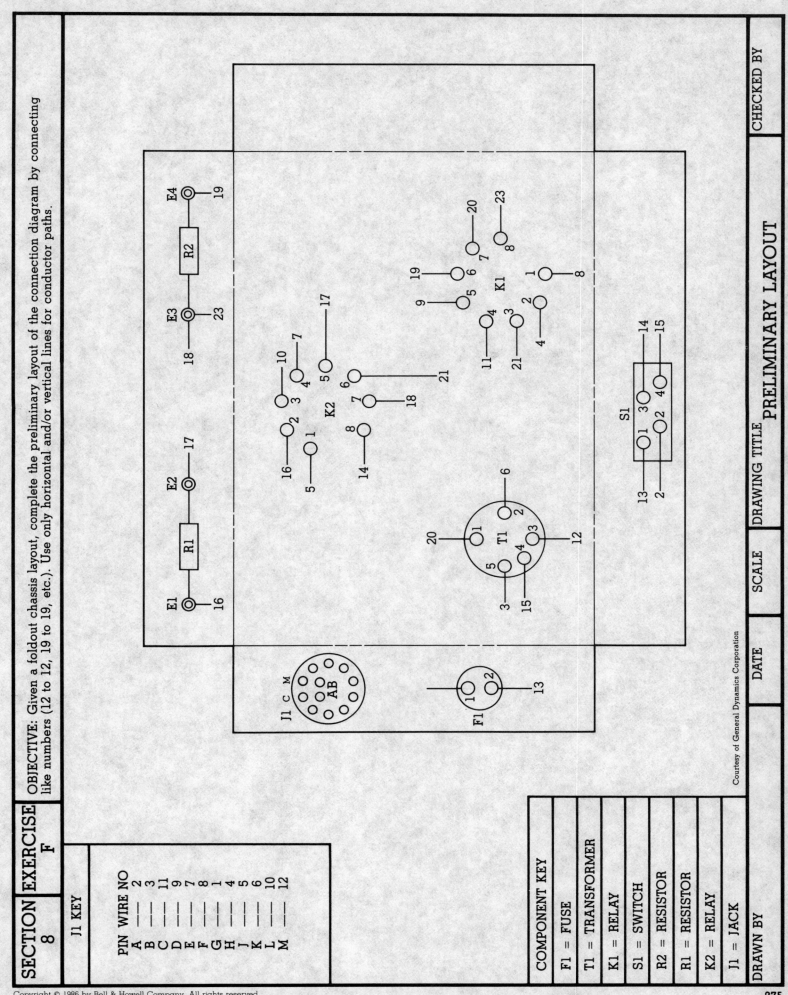

SECTION 8

EXERCISE F

OBJECTIVE: Given a foldout chassis layout, complete the preliminary layout of the connection diagram by connecting like numbers (12 to 12, 19 to 19, etc.). Use only horizontal and/or vertical lines for conductor paths.

J1 KEY

PIN	WIRE NO
A	2
B	3
C	11
D	9
E	7
F	8
G	1
H	4
J	5
K	6
L	10
M	12

COMPONENT KEY

F1 = FUSE

T1 = TRANSFORMER

K1 = RELAY

S1 = SWITCH

R2 = RESISTOR

R1 = RESISTOR

K2 = RELAY

J1 = JACK

DRAWN BY

DATE

SCALE

DRAWING TITLE

PRELIMINARY LAYOUT

CHECKED BY

Courtesy of General Dynamics Corporation

EXERCISE
G

OBJECTIVE: Using the previous exercise (8F) as a refrence produce a wiring harness diagram similar to the one shown in Figure 8.9 of the text. Label each termination. No dimensions are necessary.

DRAWN BY

DATE | SCALE | DRAWING TITLE DIAGRAM, WIRING HARNESS

CHECKED BY

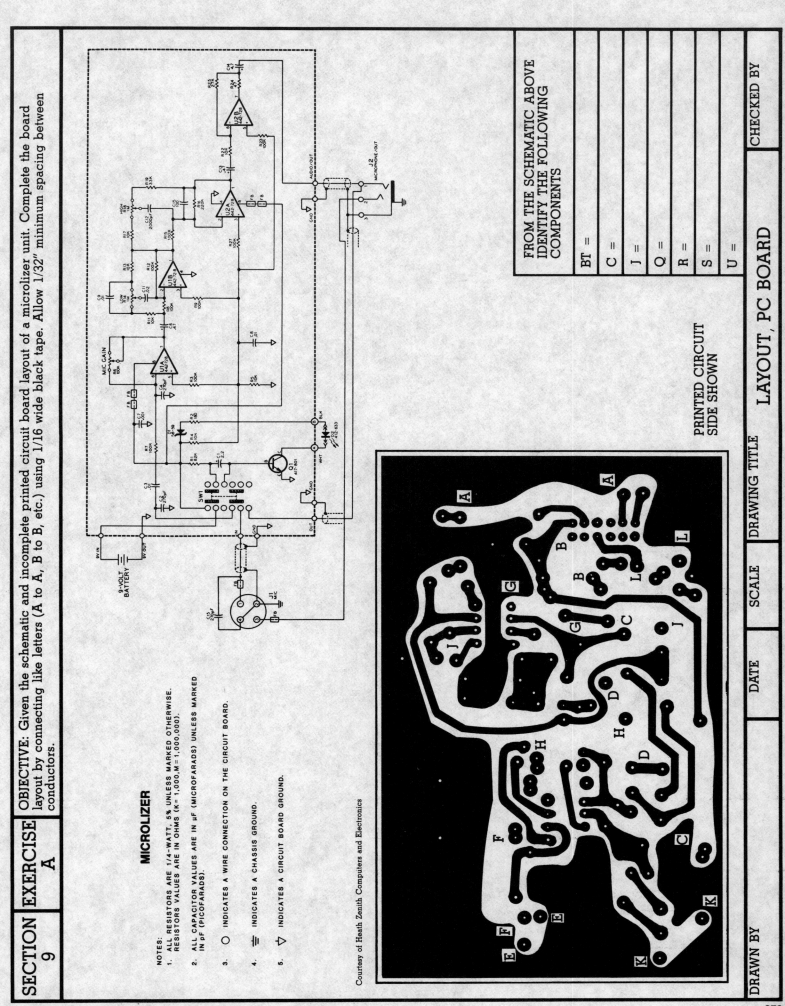

OBJECTIVE: Given the schematic and incomplete printed circuit board layout of a microlizer unit. Complete the board layout by connecting like letters (A to A, B to B, etc.) using 1/16 wide black tape. Allow 1/32" minimum spacing between conductors.

MICROLIZER

NOTES:

1. ALL RESISTORS ARE 1/4-WATT, 5% UNLESS MARKED OTHERWISE. RESISTORS VALUES ARE IN OHMS (K = 1,000, M = 1,000,000).

2. ALL CAPACITOR VALUES ARE IN μF (MICROFARADS) UNLESS MARKED IN pF (PICOFARADS).

3. ○ INDICATES A WIRE CONNECTION ON THE CIRCUIT BOARD.

4. ⏚ INDICATES A CHASSIS GROUND.

5. ▽ INDICATES A CIRCUIT BOARD GROUND.

Courtesy of Heath Zenith Computers and Electronics

FROM THE SCHEMATIC ABOVE
IDENTIFY THE FOLLOWING
COMPONENTS

BT =

C =

J =

Q =

R =

S =

U =

PRINTED CIRCUIT
SIDE SHOWN

DRAWN BY	DATE	SCALE	DRAWING TITLE	CHECKED BY
			LAYOUT, PC BOARD	

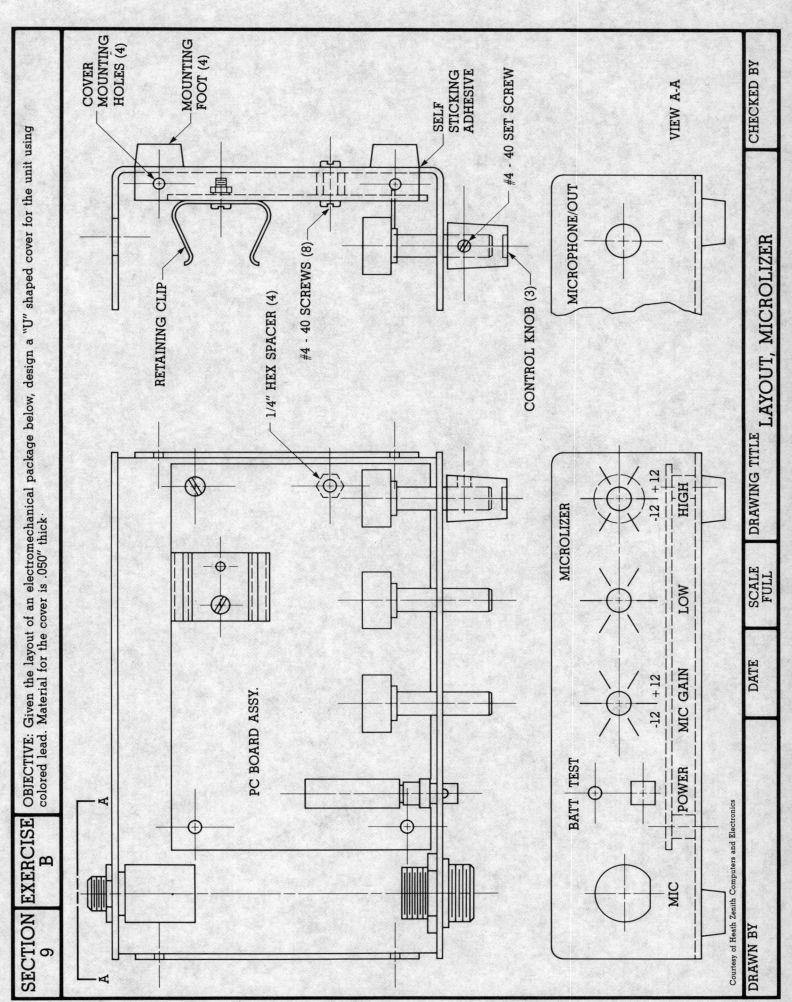

OBJECTIVE: Given the layout of an electromechanical package below, design a "U" shaped cover for the unit using colored lead. Material for the cover is .050" thick.

COVER MOUNTING HOLES (4)

MOUNTING FOOT (4)

RETAINING CLIP

1/4" HEX SPACER (4)

#4 - 40 SCREWS (8)

SELF STICKING ADHESIVE

#4 - 40 SET SCREW

CONTROL KNOB (3)

PC BOARD ASSY.

A

A

VIEW A-A

MICROPHONE/OUT

MICROLIZER

BATT TEST

-12 +12

MIC GAIN

POWER

-12 +12

LOW HIGH

MIC

DATE	SCALE FULL	DRAWING TITLE LAYOUT, MICROLIZER

CHECKED BY

DRAWN BY

OBJECTIVE: Produce a detail drawing of the U-shaped chassis cover from the layout in exercise 9B. Add dimensions. Use 16 gage (.050) aluminum material. The finish is flat black paint.

NOTES:
MATERIAL —

FINISH —

| DRAWN BY | DATE | SCALE 1/2 SIZE | DRAWING TITLE COVER, CHASSIS | CHECKED BY |

OBJECTIVE: Produce a detail drawing of the microlizer chassis from the layout in exercise 9B. Add dimensions. Use 16 gage (.050) aluminum material. The finish is flat black paint.

NOTES:
MATERIAL —

FINISH —

| DRAWN BY | DATE | SCALE 1/2 SIZE | DRAWING TITLE CHASSIS | CHECKED BY |

EXERCISE E

OBJECTIVE: Produce a detail drawing of the battery retaining clip shown on the layout in exercise 9B. Add dimensions. Material is .018 spring steel. The finish is zinc chromate.

NOTES:
MATERIAL —

FINISH —

DRAWN BY

DATE

SCALE
FULL

DRAWING TITLE
CLIP, RETAINING

CHECKED BY

SECTION 9

EXERCISE F

OBJECTIVE: Produce a detail drawing of the mounting foot shown on the layout in exercise 9B. Add dimensions. Material is high impact polymer. Color black.

NOTES:
MATERIAL —

COLOR —

DRAWN BY

CHECKED BY

DATE

SCALE
2:1

DRAWING TITLE FOOT, MOUNTING

SECTION 9	EXERCISE G	OBJECTIVE: Produce a detail drawing of a control knob identified on the layout in exercise 9B. Show dimensions. Material is polyvinylchloride (PVC). The color is black.

NOTES:
MATERIAL —

COLOR —

DRAWN BY			
	DATE	SCALE 2:1	DRAWING TITLE KNOB, CONTROL
CHECKED BY			

SECTION
9

EXERCISE
H

OBJECTIVE: Produce a marking (screen artwork) drawing of the front panel of the microlizer shown on the layout in exercise 9B. Show .093" high vertical letters. Markings are white enamel paint. Show dimensions to locate markings.

NOTE:
MARKING PAINT —

DRAWN BY

DATE

SCALE
FULL

DRAWING TITLE MARKING, FRONT PANEL

CHECKED BY

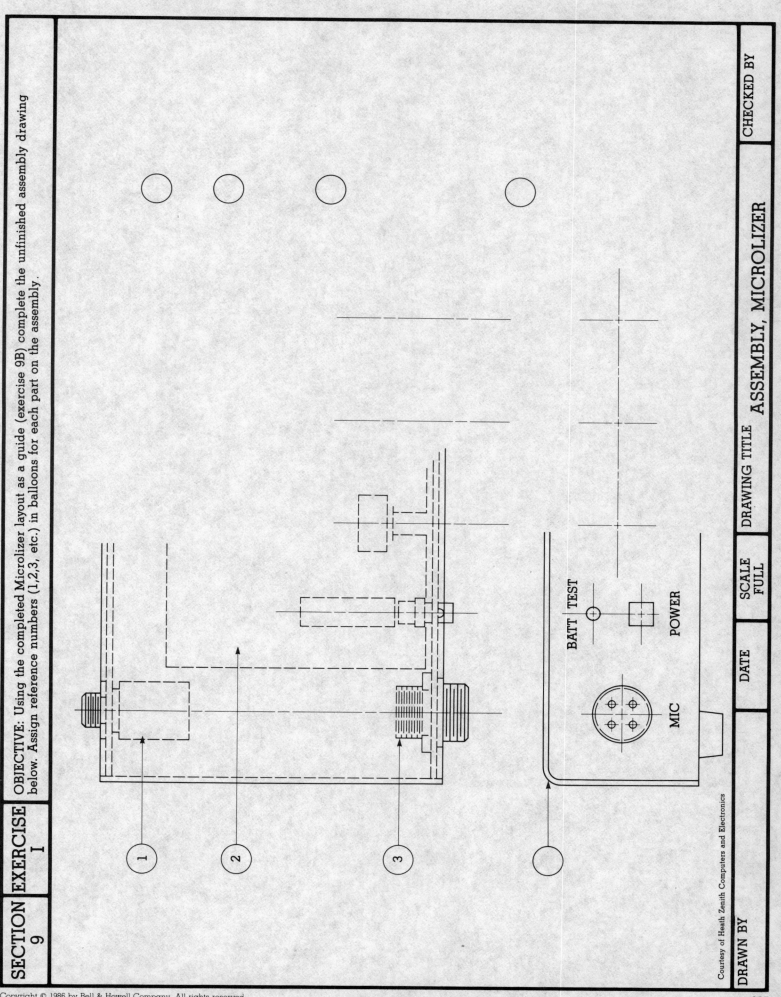

SECTION 9

EXERCISE I

OBJECTIVE: Using the completed Microlizer layout as a guide (exercise 9B) complete the unfinished assembly drawing below. Assign reference numbers (1,2,3, etc.) in balloons for each part on the assembly.

BATT TEST

POWER

MIC

DRAWN BY

DATE

SCALE
FULL

DRAWING TITLE   ASSEMBLY, MICROLIZER

CHECKED BY

SECTION 9

EXERCISE J

OBJECTIVE: Complete the list of material for the Microlizer Assembly. Assign a 14600-XXX number to each part of the assembly. Use exercise 9B and 9I for reference.

LIST OF MATERIALS

ITEM NO.	QUANTITY	PART NUMBER	DESCRIPTION
1	1	14600-001	CHASSIS

DRAWN BY	DATE	SCALE	DRAWING TITLE	CHECKED BY
			L/M, MICROLIZER	

SECTION 10.

PRACTICE EXERCISES

THE FOLLOWING EXERCISES ARE TO BE COMPLETED USING A COMPUTER-AIDED DRAFTING (CAD) SYSTEM INCLUDING ITS ASSOCIATED SOFTWARE. MAINFRAME, MINI OR MICROCOMPUTERS MAY BE USED TO DEVELOP THESE DRAWINGS AND DIAGRAMS. IN ADDITION, AN APPROPRIATE PLOTTING DEVICE TO PRODUCE A HARD COPY OF EACH EXERCISE IS RECOMMENDED.

10A. OBJECTIVE: Produce a CAD drawing of the "A" size (8½ x 11 inch) format shown below. This format is to be used as the border for all the exercises in this section. Add text information shown.

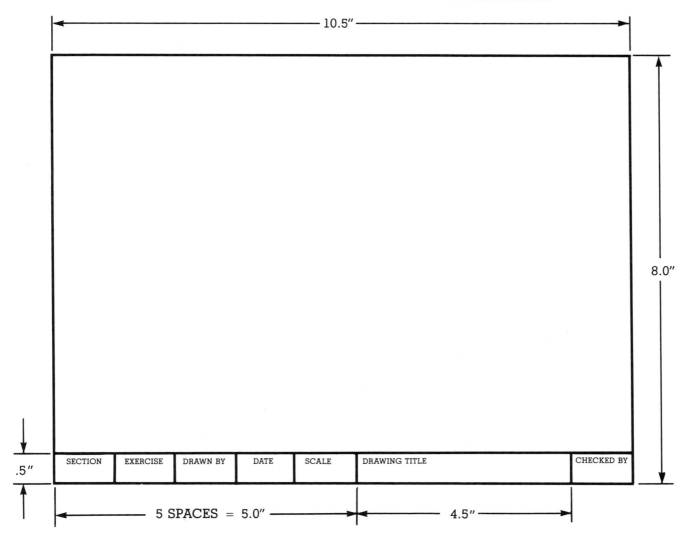

10B. OBJECTIVE: Redraw the sketch identified below. The objects should be drawn proportional to the ones shown and centered on the format. The title of the drawing is "OBJECTS." Include all lines.

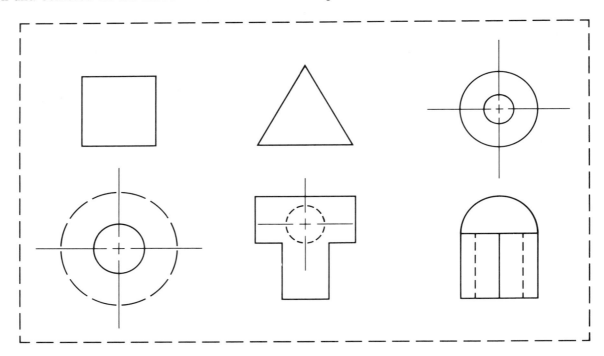

10C. OBJECTIVE: Draw the three figures shown in column A. At completion, and after approval by the instructor, change them to look like those in column B by editing, erasing, and/or adding lines. Add text information. The title of the drawing is "EDITING."

COLUMN A
FROM

COLUMN B
TO

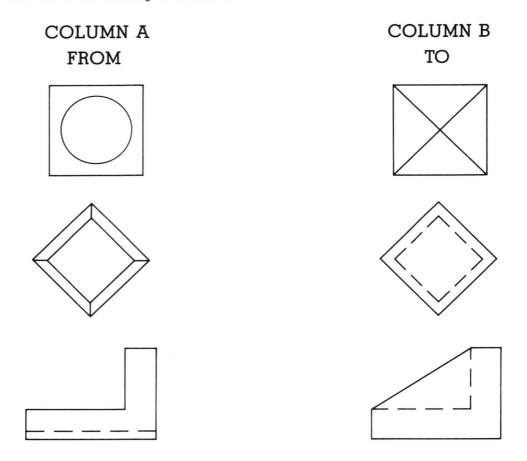

10D. OBJECTIVE: Produce blocks used on a block diagram to the ratio indicated below. Draw three of each. Include the outline as shown. The title of the drawing is "BLOCKS."

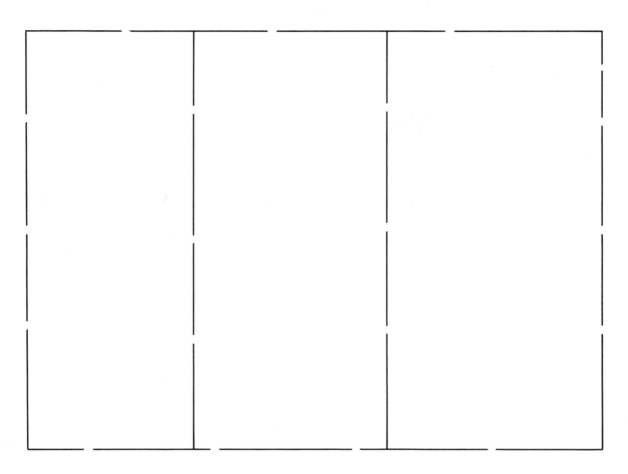

| 1:1 | 1.5:1 | 2:1 |

10E. OBJECTIVE: Produce a CAD drawing of the freehand sketch of the diagram shown. Add all text information. Draw blocks to a 1.5:1 ratio. The drawing title is "BLOCK DIAGRAM."

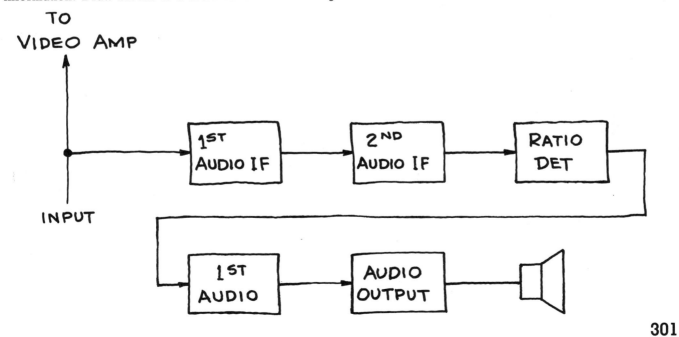

10F. OBJECTIVE: Using Appendix F, produce a CAD drawing of R1; 6.7K OHM, 1 WATT, AXIAL LEAD, WIREWOUND RESISTOR. Draw four times (4X) size. Add dimensions. The title is "OUTLINE DRAWING."

10G. OBJECTIVE: Using Table 4.1 in the text as reference, draw symbols for the following logic functions: (1) AND GATE, (2) OR GATE, (3) FLIP-FLOP, and (4) INVERTER. Draw each twice. Label and identify each device. Symbols from a logic symbol menu tablet may be used, if available. The title is "LOGIC SYMBOLS."

10H. OBJECTIVE: Given a rough sketch of a portion of a sweep generator, produce a CAD generated diagram. Include all text information. The title is "LOGIC DIAGRAM, DISTINCTIVE SHAPE."

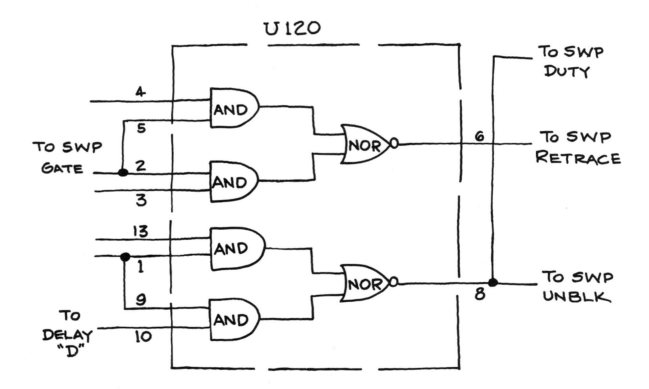

10I. OBJECTIVE: From Appendix A, draw six (6) different electronic graphics symbols. Draw each three times. Identify each set. Figures from an electronics graphics symbols menu tablet may be used, if available.

This practice of duplicating characters is called "step and repeat." The drawing title is "ELECTRONICS GRAPHICS SYMBOLS."

10J. OBJECTIVE: Reproduce the Preamplifier Circuit Schematic in exercise 5C to be a CAD generated diagram. Include all notes and component information. The title is "SCHEMATIC, PREAMPLIFIER."

10K. OBJECTIVE: Produce a schematic diagram from the freehand Counter Circuit sketch shown in exercise 5D. Include all component information. The title is "SCHEMATIC, COUNTER."

10L. OBJECTIVE: Duplicate the Taping Practice, exercise 6A. Place centers of pads and land areas on 0.10 inch grid intersections. Fill in conductor paths by using the solid, fill, or cross hatch function on the CAD system. The title is "CONDUCTOR PATHS."

10M. OBJECTIVE: Produce a component board layout, to scale, from the information provided in exercise 6C. Do not add dimensions. Calculate the required area of the board (in square inches) using the CAD system for the calculations. The title is "LAYOUT, COMPONENT BOARD."

10N. OBJECTIVE: Using three (3) different drawing layers, draw the Voltage Sensor Assembly shown in exercise 6K. On layer one (1) draw only the assembly. On layer two (2) draw the marking drawing shown in exercise 6J. Layer three (3) includes balloons, balloon numbers, leaders, and arrowheads shown in exercise 6K, which identify parts that make up the assembly. The title is "ASSEMBLY, VOLTAGE SENSOR."

10O. OBJECTIVE: Produce a CAD drawing of the rough sketch of the diagram shown below. Include all text information. The title of the drawing is "INTERCONNECTION DIAGRAM, WIRING TYPE."

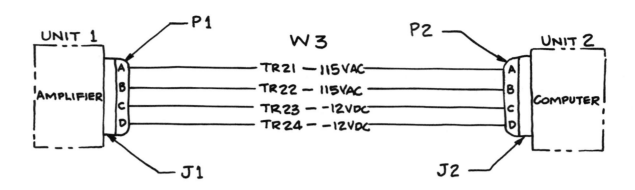

10P. OBJECTIVE: Given a sketch of the diagram identified below, redraw it with the following changes: (1) re-route wire from A1/2 to conect to TB1/4, (2) add a wire from A2/1 to connect to TB1/3. Include all text information. The title of the drawing is "CONNECTION DIAGRAM, POINT-TO-POINT."

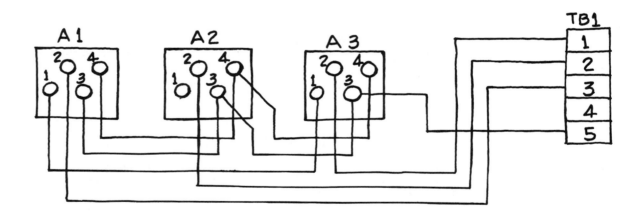

10Q. OBJECTIVE: Draw to scale, three (3) views of the bracket shown. Add all dimensions, notes, material, and finish information. The title of the drawing is "BRACKET, MOUNTING."

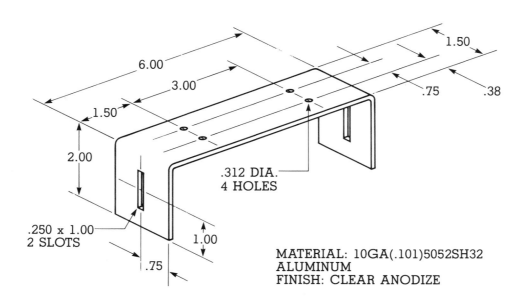

6.00
3.00
1.50
2.00
1.50
.75
.38
.312 DIA.
4 HOLES
.250 x 1.00
2 SLOTS
1.00
.75

MATERIAL: 10GA(.101)5052SH32
ALUMINUM
FINISH: CLEAR ANODIZE

10R. OBJECTIVE: Produce to scale, an isometric drawing of the part shown below. Add all dimensions, notes, material, and finish information. The title of the drawing is "COVER, CHASSIS."

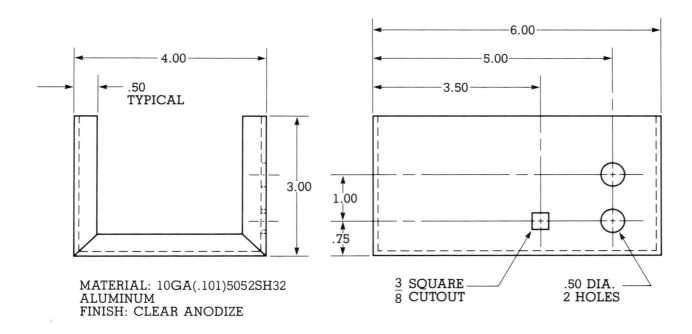

4.00
.50
TYPICAL
3.00

6.00
5.00
3.50
1.00
.75

MATERIAL: 10GA(.101)5052SH32
ALUMINUM
FINISH: CLEAR ANODIZE

$\frac{3}{8}$ SQUARE CUTOUT

.50 DIA.
2 HOLES

305

Index

Alpha characters, 118
Alphanumeric keyboard, 105, 107, 118
American National Standards Institute (ANSI), 5, 14, 26, 46, 52, 76, 83
AMES lettering guide, 7, 8
AMPLIFIER, 45
AND gate, 45, 46
ANSI. *See* American National Standards Institute
Appliqué shapes, 63, 65, 66
Arrowheads, 47
Artwork drawing, 62, 65, 66–67
Assembly, 2
Assembly drawing, 62, 67, 71–72
AutoCAD, 114–17
Auxiliary lines, 24
Auxiliary views, 13–14, 15

Back-up disk, 118
Basic logic diagram, 44
Bending, 95
Binary numbers, 44, 118
Bit, 44, 118
Block diagram, 2, 24–27, 104
    drawing procedure, 26
    lettering, 25–26, 27
    method for drawing, 26
    purpose and function, 24
Blocks, 25, 27
Board detail drawing, 62, 67, 69
Bonding, 97
Boundary markers, 66, 68
Brazing, 97
Break line, 9
Breakout, 88
Breakout numbers, 88, 89
Byte, 118

Cabinet, 95
Cable
    circular, 86
    conventional, 86
    flat, 85, 86
    ribbon, 85, 86
Cable assembly drawing, 84, 85–86
Cable form diagram, 87
Cabling type diagram, 76, 77
CAD system, 2, 104–19
    basic concept, 104
CADAM, 105–109
CADAPPLE, 112–14
Cathode-ray tube (CRT), 118
Center line, 9
Central processing unit (CPU), 105, 118
Central processor, 112
Chassis, 95, 96
    box type, 95, 96
    layout, 96
    U-shaped, 95, 96
Circular cable, 86

Class designation letters, 54
Command, 114, 118
Compass, 4, 5
Component drawing, 32
Component orientation, 62
Component sequence number, 54
Component values, 54
Computer, 105, 112, 114, 118
Computer-aided drafting system. *See* CAD system
Computer graphics industry acronyms, 104
Conductor corners, 65, 66
Conductor lengths, 65, 66
Conductor paths, 53, 54
Connecting lines, 53–54, 77
Connection diagram, 2, 82–89, 104. *See also* wiring diagram, cable assembly drawing, wiring harness diagram
Console, 95
Continuous line, point-to-point diagram, 76
Control drawing, 2, 32–40
    preparation of, 36
    purpose and function, 32
    types of, 32–35
Control files, 110
Conventional cable, 86
Coordinate, 114
Copiers, use of, 5–6
CPU, 105, 118
Crossing lines, 16, 17
CRT, 118
Cursor, 118
Curves, 5, 6
Cutouts, 97
Cutting plane line, 9

Data base, 107
Data base storage, 117
Datum-line dimensioning, 15
Design specifications, 32
Designer, 94
Desk-top computer, 105, 109
Detailed logic diagram, 44
Diagrammatic-type diagram, 76
Digitizer, 105, 112, 118
Digitizing, 118
Digitizing tablet, 114
Dimension, 14
Dimension line, 9
Dimensioning, 14–18, 88
Dimensions, 17–18
    application of, 17–18
    holes for, 18
    grouping, 17
    placement of, 17–18
    staggering of, 17
Disk drive, 109, 114, 118
Diskette, 113

Display, 110, 115, 118
Display monitor, 113
Display screen, 114
Distinctive shape logic diagram, 44
Downtime, 118
Drafter, 2, 10, 94
Drafting board, 2
Drafting chairs, 3
Drafting machines, 2, 3
Drafting station, 2, 3
Drawing materials, 5–6
Drawing sizes, 6
Drawing units, 114

Electromechanical drawings, 104
Electromechanical packaging, 2, 94–100
    method for developing, 100
Enter, 118
Entity, 114
Envelope drawing, 32, 33
Equipment, 2
Equipment enclosures, 94–95
EXCLUSIVE OR, 45
Execute, 118
Extension line, 9

Fabrication drawing, 67
Fastening methods, 97–98
Feeder lines, 82
Files, 105, 110, 118
Film, 5
Flat cable, 85–86
FLIP-FLOP, 45, 46
Floppy disk, 112, 113, 118
Flow path, 24
Font, 118
Forming, 95
Freehand lettering, 6
Freehand sketch, 10–12, 24
Function, 118
Function identification letters, 47
Function keyboard, 105, 107

Gothic style lettering, 6, 7
Graphic input devices, 116
Graphic symbols, 24, 52
Graphics display, 105, 118
Graphics display work station, 106
Graphics editor, 110
Graphics monitor, 114
Graphics tablet, 105, 118
Grip, 99, 100
Grip length, 99, 100
Group, 113
Grouping dimensions, 16

HP-EGS, 109–12
Hard copy, 106, 111, 118
Hard disk, 118
Hardware, 105, 119

Hidden line, 9
Highway type diagram, 82–83
Host, 119

Inclined lettering, 6
Information flow, 24, 25, 52
Input, 104
Input device, 107, 112, 116
Interactive, 104, 105, 113
Interconnection diagram, 2, 76–78, 104
  layout, 77
  method for drawing, 77–78
  purpose and function, 76
  types of, 76
INVERTER, 45
Isometric freehand sketch, 11
Isometric projection, 12

Joystick, 112, 119

Keyboard, 112, 114

Land area, 63
Layer, 113, 116
Lead holders, 10
Leader line, 9
Leads, 10
Lettering, 6
Lettering aids, 5, 7–8
Level, 113
Light pen, 107, 116, 119
Line convention, 9–10, 24–25, 26, 52
Line weight, 52, 78, 84, 85
List of materials, 65, 104
Logic diagram, 2, 44–48, 104
  method for drawing, 47
  purpose and function, 44
  types of, 44
Logic functions, 44–46
  description, 45
  identification letter combination, 46
  input/output table, 45
Logic states, 44
Logic symbol presentation techniques, 44, 46
Logic symbols, 44, 45

Macro files, 110
Mainframe, 105, 119
Marking drawing, 62, 67, 70
Master drawing, 62, 67
Master pattern drawing, 62, 65
Mechanical drawings, 104
Mechanical pencil, 10
Memory, 109, 112, 119
Memory device, 105
Menu, 109, 115
  graphics tablet, 110
  screen, 110

  tablet, 117, 119
Menu files, 110
Message, 119
Message files, 110
Microcomputer, 105, 119
Microcomputer-based system, 112–17
Military Standard (MIL-STD), 6, 26, 44, 46, 97
MIL-STD. *See* Military Standard
Minicomputer based systems, 105–12
Modem, 119
Monitor, 112, 119
Mouse, 116
Multiview freehand sketch, 11

NAND gate, 45
Nonlogic functions, 44
NOR gate, 45
Notes, 6, 54, 83

Objects, 113, 114
One-view drawings, 12–13
OR gate, 45
Orthographic projection, 12–14
Outline, 9
Outline drawing, 32
Output, 104

Packaging, 94
Pan, 113, 119
Panel, 95
Parallel straightedge, 3
Part, 2
Pen carousel, 112
Performance data, 32
Peripheral, 119
Personal computer, 113
Perspective projection, 12
Phantom line, 9
Photographer's reduction dimension, 66, 68
Plot, 119
Plotter, 105, 109, 111, 112, 113, 117, 119
Plotting, 117
Point-to-point diagram, 82
Pointing devices, 116
Primitive element, 111, 114
Principal view, 17
Printed circuit, 62
Printed circuit board, 2, 62–72
  layout, 62–65, 104
  layout review, 63, 65
  purpose and function, 62
Printed wiring board. *See* Printed circuit board
Process files, 110
Processor, 114
Projection drawing, 12
Printer, 109

Program function key, 109
Programs, 105
Prompt, 119
Property, 114
Puck, 105, 119
Purchased part drawing, 32

Rack, 94
Radii dimensioning, 18
Raster, 119
Rectangular shape logic diagram, 44
Reference designation location, 54
Reference designations, 54, 55
Reference table, 3
Registration marks, 65–66, 67, 68
Remote station, 119
Repeatability, 111
Resolution, 111, 115, 119
Ribbon cable, 85–86
Riveting, 97
Running list, 88, 104

Schematic diagram, 2, 52–55, 64, 104
  conductor paths for, 52
  symbols for, 52
  method for drawing, 55
  purpose and function, 52
Screen, 113, 119
Screen menu, 114, 116
Screw chart, 97
Screw length chart, 99
Screws
  clearance holes, 97–99
  length selection, 99–100
Section line, 9
Sheets, 113
Signal flow, 47
Signal path, 47
Silkscreen drawing, 62, 67
Single stroke letter, 6
Six-view orthographic projection, 12, 13
Software, 105, 110, 112, 119
Soldering, 97
Source control drawing, 32, 35, 36
Specification control drawing, 32, 34, 37–40
Staking, 97
Storage facility, 105
Stylus, 105, 119
Subassembly, 2
System, 2

T-square, 3
Tablet, 107, 116
Tabular type diagram, 76, 83
Tagging lines, 46
Target, 66
Technical graphics, 2, 118
Template, 4–5, 7–8, 46, 63

Terminal, 119
Terminal area, 65, 67
Third-angle orthographic projection, 12
Three-view drawings, 13, 14, 16
Tools, 4
Touch pen, 116
Triangles, 4
Trim marks, 66
TRI-STATE BUFFER, 44, 45
Trunkline wiring diagram, 82
Turnkey, 105, 119
Two-dimensional, 110
Two-dimensional drafting system, 105

Two-view drawings, 13, 14

Unidirectional dimensioning, 15
Unit, 2
Uppercase letter, 6

Vellum, 5
Vertical lettering, 6
Viewing plane line, 9
Visible line, 9

Welding, 97
Winchester disk, 118, 119

Window, 113
Wire data list, 83
Wire list, 88
Wiring diagram, 76, 77, 82–85
    method for drawing, 84
    purpose and function, 82
    types of, 82–83
Wiring harness diagram, 87–89

Zoom, 113